Rand McNally
Deluxe Road Atlas & Travel Guide

S0-DMY-656

Contents

How to Use This Book

Today's leisure traveler faces a wealth of choices when planning a trip—not only what the destination is to be, but how best to get there, where to stay, which of countless sightseeing opportunities to engage in. Many travelers also want to plan their trips around recreational interests such as a theme park visit, fishing, or camping.

Rand McNally's *Deluxe Road Atlas & Travel Guide* brings together the details, facts, and travel information necessary for the planning and execution of a successful, enjoyable leisure trip. The Travel and Recreation Guide provides pertinent information in an easy-to-use format, valuable both in the initial trip planning stages and once the destination has been reached. The Major Cities Guide relates important facts and describes the amenities of principal cities in the United States and Canada, plus Mexico City. Together with the accurate and detailed city maps contained in this section, and those of the United States, Canada, and Mexico contained in the Atlas section, these features create an all-in-one guide and atlas that takes the leisure traveler from the planning stage to the destination—and back home again.

Travel and Recreation Guide

Arranged in easy-to-use charts, this guide provides ready access to recreation and travel opportunities and resources around the country. The following topics are covered in this section.

United States Mileage Chart.   An aid to estimating driving time, this chart provides the mileage between a host of major United States cities.

Resorts.   This listing of selected resorts includes address/phone and details regarding locale and range of facilities.

Theme Parks.   Hours of operation and other pertinent information on 30 theme parks around the U.S. and Canada.

Fishing/Hunting.   License fee information is provided for all 50 states.

Special Events.   A selected list of special events taking place throughout the United States and Canada.

National Parks.   Outdoor sports enthusiasts, campers, nature lovers, and history buffs can use this chart to determine activities, facilities, and accommodations available at over 200 National Park areas, monuments, and historic sites.

Tourism Information and Auto/RV Laws.   Keep abreast of state regulations pertaining to trailer towing, child restraints, seat belts, and more.

Canada Travel Information.   Canadian-U.S. border crossing regulations and information about fishing and hunting in both countries is summarized here.

Road Condition "Hot Lines."   Current status for road conditions and construction projects can be obtained by calling these "hot line" numbers.

Intercity Toll Road Information.   An aid to trip planning, this chart gives the location, mileage, and cost of using toll roads in the United States and Mexico.

Car Rental Toll-free Numbers. Locate your rental car before starting your trip, using these convenient toll-free numbers.

Hotel/Motel Toll-free Numbers. These toll-free telephone numbers for more than 50 hotel and motel chains can save both time and money.

Clear Channel Radio Stations.   Specially organized for quick-reference use while driving; dial one of these sta-

tions and enjoy clear reception over an extended range.

Climate.   In-depth climate information is provided in this chart for all 50 states plus the District of Columbia.

Mileage and Driving Time Map. Mileages and driving times between major places in the United States, Canada, and Mexico are provided in a handy map format.

Maps

The atlas section includes a United States map and maps of each of the 50 states, Canada, the Canadian provinces, and Mexico.

Valuable as aids to routing your trip, these maps are also useful for pinpointing sites referred to in the Recreation and Travel section charts, and interpreting that section's mileage and climate information.

Major Cities Guide

Information is provided for more than 65 major cities in the United States, Canada, and Mexico, including:

Selected Hotels, Restaurants and Attractions.   Listed for each city, these hotels, restaurants, and attractions have been chosen for their broad appeal to the traveling public.

City Maps.   Large-scale maps are included for each city to aid in orienting the first-time visitor, and refreshing the memory of the returning traveler.

Airport Maps.   Flyers will appreciate the detailed airport maps provided for 27 of the largest United States cities.

Information Sources.   The city listings also include the address/phone for local convention & visitors bureaus, offices of tourism, and chambers of commerce.

Reproducing or recording maps, tables, text, or any other material which appears in this publication by photocopying, by electronic storage and retrieval, or by any other means is prohibited.

Information included in this publication has been checked for accuracy prior to publication. Since changes do occur, the publisher cannot be responsible for any variation from the information printed.

Copyright © 1993 by Rand McNally & Company
Made in U.S.A. All rights reserved
Proof of Purchase RMC 93 DRATG

Travel and Recreation Guide

The 14 charts and map included in this guide are designed to assist in both the planning and execution of your trip; information provided covers sightseeing, recreation, places to stay, and on-the-road service information.

Look for the following specifics in the Travel and Recreation charts.

United States Mileage Chart. As an aid to estimating driving time, this chart provides the mileage between a host of major United States cities.

Resorts. Details on selected resorts in the U.S. and Canada: address/phone, waterfront locales, and the range of available facilities from pools to exercise rooms to sports to children's programs/playgrounds.

Theme Parks. This chart covers 30 major theme parks — noting when the parks are open; the types of tickets available; special facilities such as handicap access, picnic; and address/phone numbers.

Fishing/Hunting. Fishing and hunting license fees are detailed in this chart. For further information, the phone numbers of state licensing agencies are noted.

Special Events. A selected number of special events ranging from Mardi Gras celebrations to a balloon fiesta, rodeos to winter carnivals. Included is information as to the specific type of celebration involved, the approximate date(s) held, and addresses/phone numbers to contact for further information.

National Parks. Activities, facilities and accommodations at over 200 National Park areas, monuments, and historic sites. Activities cover various winter and water sports, horseback riding, hiking, guided tours. Facility information includes handicap access, food service, museums, bathhouses. Accommodation information covers campgrounds, motels/lodges, and cabins.

Tourism Information & Auto/RV Laws. This chart provides information on state regulations pertaining to trailer towing, seat belts, child restraints, and more. Also, phone numbers for the individual state tourism departments are listed.

Canada Travel Information. Border crossing information for Canadians and Americans is provided as well as information about hunting and fishing regulations in the U.S. and Canada. Sources to contact for information in each Canadian province are listed.

Road Condition "Hot Lines." Use these "hot line" numbers to obtain up-to-date information on road conditions and construction projects.

Intercity Toll Road Information. United States and Mexico; includes road names, location, mileage, and tolls for both autos alone and those pulling a two-axle trailer.

Hotel/Motel and Car Rental Toll-free Numbers. A list of toll-free reservation numbers for lodging and auto rental firms in the continental U.S. and Canada.

Clear Channel Radio Stations. A listing by state and province and call numbers of AM stations designated to render service over an extended area.

Climate. In-depth climate information for all 50 states plus the District of Columbia, including maximum and minimum normal daily temperatures, total precipitation, and total days with precipitation for all four seasons of the year.

Mileage and Driving Time Map. Mileages and driving times between places and major junctions are provided. Driving times take into account topography, congested urban areas and speed limits imposed.

United States Mileage Chart

89-7501 ©Rand McNally & Co.	Atlanta, Ga.	Boston, Mass.	Cheyenne, Wyo.	Chicago, Ill.	Cincinnati, Ohio	Cleveland, Ohio	Dallas, Texas	Denver, Colo.	Des Moines, Iowa	Detroit, Mich.	Houston, Texas	Indianapolis, Ind.	Kansas City, Mo.	Los Angeles, Calif.	Louisville, Ky.	Memphis, Tenn.	Milwaukee, Wis.	Minneapolis, Minn.	New Orleans, La.	New York, N.Y.	Omaha, Nebr.	Philadelphia, Pa.	Pittsburgh, Pa.	Portland, Oreg.	St. Louis, Mo.	Salt Lake City, Utah	San Francisco, Calif.	Seattle, Wash.	Toledo, Ohio	Tulsa, Okla.	Washington, D.C.	Wichita, Kans.
Albuquerque, N. Mex.	1381	2172	517	1281	1372	1560	638	417	977	1525	834	1266	782	807	1301	1010	1319	1190	1134	1979	858	1899	1619	1371	1038	604	1115	1440	1469	645	1824	593
Amarillo, Texas	1097	1897	511	1043	1096	1285	358	423	742	1269	596	991	547	1091	1019	726	1084	975	850	1704	643	1624	1344	1655	756	888	1399	1724	1210	361	1549	350
Atlanta, Ga.		1037	1442	674	440	672	795	1398	870	699	789	493	798	2182	382	371	761	1068	479	841	986	741	687	2601	541	1878	2496	2618	640	772	608	903
Austin, Texas	919	1911	994	1110	1083	1327	195	906	877	1315	186	1037	682	1374	982	615	1184	1129	517	1715	837	1615	1367	2069	823	1302	1748	2138	1256	450	1482	548
Baltimore, Md.	645	392	1608	668	497	343	1356	1621	981	503	1412	563	1048	2636	598	904	755	1073	1115	196	1113	102	218	2751	798	2044	2796	2681	444	1194	45	1244
Birmingham, Ala.	153	1165	1347	642	465	709	645	1286	787	724	639	475	697	2032	364	249	728	1006	342	969	898	869	741	2505	465	1781	2366	2535	665	647	736	778
Bismarck, N. Dak.	1495	1794	572	831	1118	1166	1141	671	670	1097	1384	1012	777	1617	1123	1228	758	427	1583	1633	581	1569	1283	1265	979	916	1604	1195	1063	958	1502	776
Boise, Idaho	2174	2639	732	1683	1906	2011	1582	811	1359	1942	1778	1800	1382	849	1893	1832	1692	1405	2078	2478	1227	2410	2122	432	1665	340	658	501	1908	1486	2343	1312
Boston, Mass.	1037		1907	994	840	628	1748	1949	1280	695	1804	906	1391	2979	941	1296	1050	1368	1507	222	1412	296	574	3046	1141	2343	3095	2976	739	1537	429	1587
Buffalo, N. Y.	859	446	1466	522	431	191	1346	1508	839	253	1460	481	966	2554	532	899	609	927	1217	372	971	353	224	2605	750	1937	2654	2535	298	1112	356	1198
Charleston, S. Car.	289	929	1722	877	603	730	1072	1678	1150	842	1054	696	1078	2459	591	660	964	1282	720	733	1266	633	666	2881	821	2158	2785	2890	783	1061	500	1192
Cheyenne, Wyo.	1442	1907		954	1190	1279	869	100	627	1211	1107	1068	650	1137	1161	1101	987	788	1361	1746	495	1678	1390	1159	901	436	1188	1228	1176	765	1611	583
Chicago, Ill.	674	994	954		287	335	917	996	327	266	1067	185	499	2054	292	530	90	405	912	802	459	738	452	2083	289	1390	2142	2013	244	683	671	696
Cincinnati, Ohio	440	840	1190	287		244	920	1173	589	269	1029	111	591	2179	103	468	374	692	786	647	693	567	295	2333	340	1610	2362	2300	210	736	481	787
Cleveland, Ohio	672	628	1279	335	244		1159	1321	652	170	1273	294	779	2367	345	712	422	740	1030	473	784	413	129	2418	529	1715	2467	2348	116	925	346	975
Columbus, Ohio	533	735	1235	308	111	144	1028	1229	618	192	1137	178	656	2244	209	576	395	713	894	542	750	462	187	2391	406	1671	2423	2321	144	802	427	852
Dallas, Texas	795	1748	869	917	920	1159		781	684	1143	246	865	489	1387	819	452	991	936	496	1552	644	1452	1204	2009	630	1242	1753	2078	1084	257	1319	365
Denver, Colo.	1398	1949	100	996	1173	1321	781		669	1253	1019	1058	606	1059	1120	1040	1029	920	1273	1771	537	1691	1411	1238	857	504	1235	1307	1218	681	1616	509
Des Moines, Iowa	870	1280	627	327	589	652	684	669		584	905	465	194	1727	566	599	361	252	978	1119	135	1051	763	1786	333	1063	1815	1749	549	443	984	392
Detroit, Mich.	738	834	1211	279	269	170	1143	1253	584		1265	278	743	2311	360	757	353	671	1045	641	716	573	287	2349	513	1647	2399	2279	62	909	506	940
Duluth, Minn.	1139	1428	918	465	752	800	1086	994	402	707	1307	646	597	2016	757	943	392	156	1331	1267	510	1203	917	1705	662	1315	2044	1635	697	845	1136	794
El Paso, Texas	1415	2316	754	1430	1515	1704	620	654	1126	1674	748	1410	931	790	1438	1072	1468	1339	1098	2123	1007	2043	1763	1635	1175	868	1164	1704	1618	780	1939	742
Flagstaff, Ariz.	1704	2495	757	1604	1695	1883	961	657	1300	1848	1157	1589	1105	484	1624	1333	1642	1481	1457	2302	1171	2222	1942	1241	1361	511	792	1347	1792	968	2147	916
Fort Wayne, Ind.	593	825	1093	170	153	197	983	1135	466	160	1105	132	586	2175	216	553	243	561	901	662	598	585	362	2239	353	1529	2281	2169	134	749	518	783
Fort Worth, Texas	826	1779	845	941	951	1183	28	757	708	1167	262	889	513	1356	850	483	1015	960	527	1583	655	1483	1235	1978	654	1211	1722	2047	1108	279	1350	359
Harrisburg, Pa.	700	373	1579	639	468	314	1383	1592	952	474	1439	534	1019	2607	569	931	726	1044	1142	198	1084	114	205	2722	769	2015	2767	2652	415	1165	133	1215
Helena, Mont.	2030	2388	685	1425	1712	1760	1554	781	1161	1691	1792	1606	1251	1190	1717	1702	1352	1020	2033	2227	1050	2163	1877	658	1493	477	1098	588	1657	1416	2096	1234
Houston, Texas	789	1804	1107	1067	1029	1273	246	1019	905	1265		987	710	1538	928	561	1142	1157	356	1608	865	1508	1313	2205	839	1438	1912	2274	1206	478	1375	608
Indianapolis, Ind.	493	906	1068	185	110	294	865	1058	465	278	987		485	2073	114	435	268	586	796	713	587	633	353	2227	246	1504	2256	2194	219	631	558	681
Jackson, Miss.	391	1406	1257	742	655	899	404	1169	809	914	406	646	644	1791	554	213	824	1036	180	1210	845	1110	939	2401	495	1646	2157	2470	855	527	977	708
Jacksonville, Fla.	306	1155	1748	980	746	915	990	1704	1176	1003	889	799	1104	2377	688	674	1067	1374	555	959	1292	859	851	2907	847	2184	2743	2924	944	1075	726	1206
Kansas City, Mo.	798	1391	650	499	591	779	489	600	195	743	710	485		1589	520	451	537	447	806	1198	201	1118	838	1809	256	1086	1835	1839	687	248	1043	197
Knoxville, Tenn.	219	911	1372	527	253	485	837	1328	800	512	893	346	728	2202	241	387	614	932	596	715	916	615	511	2531	471	1808	2510	2540	453	786	482	871
Las Vegas, Nev.	1964	2725	855	1772	1941	2097	1221	777	1445	2029	1417	1835	1365	282	1884	1593	1805	1607	1717	2548	1313	2468	2188	981	1621	433	564	1152	1994	1228	2393	1176
Lexington, Ky.	362	896	1233	352	82	317	861	1192	636	337	970	171	592	2180	74	409	439	757	727	703	758	623	343	2392	335	1669	2421	2365	278	731	514	782
Little Rock, Ark.	509	1434	1035	640	606	850	314	947	561	838	427	560	389	1687	505	137	715	813	418	1238	590	1138	890	2179	352	1438	1995	2228	779	271	1005	453
Los Angeles, Calif.	2182	2979	1137	2054	2179	2367	1387	1059	1727	2311	1538	2073	1589		2108	1817	2087	1889	1883	2786	1595	2706	2426	959	1845	715	379	1131	2276	1452	2631	1400
Louisville, Ky.	382	941	1161	292	103	345	819	1120	566	360	928	114	520	2108		367	379	697	685	748	687	668	388	2320	258	1597	2349	2305	301	659	582	710
Mackinaw City, Mich.	935	*916	1291	387	495	439	1281	1341	673	284	1427	460	864	2392	562	880	368	508	1247	906	805	842	556	2128	651	1691	2443	2058	328	1047	775	1061
Madison, Wis.	812	1103	912	144	427	475	968	954	286	406	1137	321	483	2012	432	622	77	272	1006	942	418	878	592	1950	358	1348	2100	1880	372	727	811	676
Memphis, Tenn.	371	1296	1101	530	468	712	452	1040	599	713	561	435	451	1817	367		612	826	390	1100	652	1000	752	2259	285	1535	2125	2290	654	401	867	532
Miami, Fla.	655	1504	2097	1329	1095	1264	1300	2037	1525	1352	1190	1148	1448	2687	1037	997	1416	1723	856	1308	1641	1208	1200	3256	1196	2532	3053	3273	1293	1398	1075	1529
Milwaukee, Wis.	761	1050	987	90	374	422	991	1029	361	353	1142	268	537	2087	379	612		332	994	889	493	825	539	2010	363	1423	2175	1940	319	757	758	734
Minneapolis, Minn.	1068	1368	788	405	692	740	936	920	252	671	1157	586	447	1889	697	826	332		1214	1207	380	1143	857	1678	552	1312	1940	1608	637	695	1076	644
Mobile, Ala.	335	1372	1443	851	706	950	590	1355	954	965	478	716	812	1977	605	363	933	1173	143	1176	1013	1076	982	2587	632	1832	2343	2651	906	713	943	893
Montreal, Que.	1181	318	1773	828	805	561	1705	1815	1146	562	1827	840	1305	2873	906	1273	915	1163	1591	378	1278	449	583	2900	1101	2225	2961	2685	621	1471	579	1502
Nashville, Tenn.	242	1088	1200	446	269	513	660	1156	628	528	769	279	556	2025	175	210	532	826	517	892	744	792	553	2359	299	1636	2333	2376	469	609	659	699
New Orleans, La.	479	1507	1361	912	786	1030	496	1273	978	1045	356	796	806	1883	685	390	994	1214		1311	1007	1211	1070	2505	673	1738	2249	2574	986	647	1078	816
New York, N. Y.	841	222	1746	802	647	473	1552	1771	1119	637	1608	713	1198	2786	748	1100	889	1207	1311		1251	101	368	2885	948	2182	2934	2815	578	1344	233	1394
Norfolk, Va.	540	558	1764	831	604	508	1329	1758	1141	666	1328	700	1162	2694	642	877	918	1236	1019	362	1273	272	384	2914	905	2200	2952	2844	607	1278	188	1352
Oklahoma City, Okla.	839	1641	697	787	840	1029	209	609	544	1013	449	735	349	1349	763	468	861	796	668	1448	455	1368	1088	1841	500	1100	1657	1910	954	105	1293	157
Omaha, Nebr.	986	1412	495	459	693	784	644	537	135	716	865	587	201	1595	687	652	493	380	1007	1251		1183	895	1654	449	931	1683	1638	681	435	1116	298
Orlando, Fla.	435	1294	1876	1109	875	1054	1078	1815	1305	1134	968	928	1226	2465	817	775	1196	1503	634	1098	1421	998	990	3034	976	2310	2831	3053	1075	1176	865	1307
Philadelphia, Pa.	741	296	1678	738	567	413	1452	1691	1051	573	1508	633	1118	2706	668	1000	825	1143	1211	101	1183		288	2821	868	2114	2866	2751	514	1264	143	1314
Phoenix, Ariz.	1793	2604	892	1713	1804	1992	998	792	1409	1957	1149	1698	1214	389	1733	1442	1751	1616	1494	2411	1290	2331	2051	1266	1470	648	763	1437	1901	1077	2256	1025
Pierre, S. Dak.	1361	1726	434	763	1050	1098	943	518	492	1029	1186	944	592	1524	1055	1043	690	394	1394	1565	391	1501	1215	1353	824	823	1575	1283	995	760	1434	578
Pittsburgh, Pa.	687	574	1390	452	295	129	1204	1411	763	287	1313	353	838	2426	388	752	539	857	1070	368	895	288		2535	588	1826	2578	2465	228	984	259	1034
Portland, Me.	1139	115	1986	1042	942	707	1850	2028	1359	775	1906	1001	1486	3074	1043	1398	1129	1447	1609	308	1491	398	663	3125	1236	2422	3174	3055	818	1632	531	1682
Portland, Ore.	2601	3046	1159	2083	2333	2418	2009	1238	1786	2349	2205	2227	1809	959	2320	2259	2010	1678	2505	2885	1654	2821	2535		2060	767	636	174	2315	1913	2754	1739
Raleigh, N. Car.	372	685	1695	784	534	561	1166	1661	1092	683	1160	631	1087	2545	541	728	871	1189	851	489	1214	389	445	2854	831	2131	2853	2797	624	1129	256	1215
Rapid City, S. Dak.	1487	1859	313	896	1177	1231	1050	394	618	1162	1288	1071	708	1363	1182	1159	840	565	1507	1698	507	1634	1348	1204	950	662	1414	1134	1128	873	1567	691
Reno, Nev.	2374	2866	959	1913	2133	2238	1668	1011	1586	2170	1864	2027	1609	469	2120	2003	1946	1711	2164	2705	1454	2637	2349	538	1860	523	229	754	2135	1640	2570	1471
Richmond, Va.	510	535	1674	748	514	425	1266	1668	1058	583	1298	610	1072	2631	552	814	835	1153	989	339	1190	239	301	2831	815	2110	2862	2761	524	1211	105	1262
St. Louis, Mo.	541	1141	901	289	340	529	630	857	333	513	839	246	256	1845	258	285	363	552	673	948	449	868	588	2060		1337	2089	2081	454	396	793	447
Salt Lake City, Utah	1878	2343	436	1390	1610	1715	1242	504	1063	1647	1438	1504	1086	715	1597	1535	1423	1312	1738	2182	931	2114	1826	767	1337		752	836	1612	1172	2047	1003
San Antonio, Texas	983	1988	1027	1187	1160	1404	270	939	954	1392	197	1114	759	1363	1059	692	1261	1206	550	1792	914	1692	1444	2086	900	1319	1737	2155	1333	527	1559	625
San Diego, Calif.	2126	2955	1186	2064	2155	2343	1331	1108	1760	2308	1482	2049	1565	124	2084	1783	2102	1938	1827	2762	1641	2682	2402	1084	1821	764	504	1256	2252	1428	2607	1376
San Francisco, Calif.	2496	3095	1188	2142	2362	2467	1753	1235	1815	2399	1912	2256	1835	379	2349	2125	2175	1940	2249	2934	1683	2866	2578	636	2089	752		808	2364	1760	2799	1695
Seattle, Wash.	2618	2976	1228	2013	2300	2348	2078	1307	1749	2279	2274	2194	1839	1131	2305	2290	1940	1608	2574	2815	1638	2751	2465	174	2081	836	808		2245	1982	2684	1808
Spokane, Wash.	2340	2698	995	1735	2022	2070	1864	1089	1471	2001	2102	1916	1561	1205	2027	2012	1662	1330	2343	2537	1360	2473	2187	348	1803	712	882	278	1967	1726	2406	1544
Springfield, Ill.	592	1099	888	189	299	473	728	865	291	433	879	212	310	1899	275	370	263	480	758	906	412	826	546	2047	97	1324	2076	2039	377	494	751	507
Springfield, Mo.	652	1353	820	499	552	741	419	759	383	725	629	447	192	1636	475	281	573	594	636	1160	371	1080	800	1978	210	1254	1944	2009	666	183	1005	287
Toledo, Ohio	640	739	1176	244	210	116	1084	1218	549	62	1206	219	687	2276	301	654	319	637	986	578	681	514	228	2315	454	1612	2364	2245		850	447	884
Topeka, Kans.	863	1456	595	562	656	844	480	535	255	806	701	550	65	1531	585	508	598	508	863	1263	165	1183	903	1751	322	1028	1770	1802	750	223	1108	139
Tulsa, Okla.	772	1537	765	683	736	925	257	681	443	909	478	631	248	1452	659	401	757	695	647	1344	435	1264	984	1913	396	1172	1760	1982	850		1189	174
Washington, D. C.	608	429	1611	671	481	346	1319	1616	984	506	1375	558	1043	2631	582	867	758	1076	1078	233	1116	143	259	2754	793	2047	2799	2684	447	1189		1239
Wichita, Kans.	903	1587	583	696	787	975	365	509	392	940	608	681	190	1400	710	532	734	644	816	1394	298	1314	1034	1739	447	1003	1695	1808	884	174	1239	

Mileages Copyright ©1987 by Rand McNally—TDM, Inc.

Resorts

Resort	Address/Telephone	Children's Programs	Exercise	Skiing	Horseback Riding	Tennis	Golf	Pool	Fishing	Boating	Waterfront
EASTERN U.S.											
The Balsams Grand Resort Hotel	Dixville Notch, NH 03576; 603/255-3400, 1-800/255-0800 (NH), 1-800/255-0600 (U.S. & Canada)		●	●	●	●	●		●		●
Boca Raton Resort & Club	501 E. Camino Real, Boca Raton, FL 33432; 407/395-3000, 1-800/327-0101	●	●	●	●	●				●	●
The Breakers	1 S. County Rd., Palm Beach, FL 33480; 407/655-6611, 1-800/833-3141	●				●	●	●		●	●
Le Château Montebello	392 Notre Dame, Montebello, Québec J0V1L0; 819/423-6341, 1-800/268-9411 (Can.), 1-800/828-7447 (U.S.)	●	●	●	●	●	●	●	●		●
The Cloister	Sea Island, GA 31561; 912/638-3611, 1-800/732-4752	●	●	●	●	●				●	●
Doral Resort & Country Club	4400 N.W. 87th Ave., Miami, FL 33178-2192; 305/592-2000, 1-800/367-2826 (FL), 1-800/327-6334 (U.S.)					●	●	●		●	●
Grand Traverse Resort	6300 U.S. 31N, Acme, MI 49610-0404; 616/938-2100, 1-800/748-0303 (U.S.), 1-800/678-1308 (Can.)	●	●	●	●	●	●			●	●
The Greenbrier	U.S. 60 West, White Sulphur Springs, WV 24986; 304/536-1110, 1-800/624-6070					●	●	●	●	●	●
Grove Park Inn & Country Club	290 Macon Ave., Asheville, NC 28804; 704/252-2711, 1-800/438-5800					●	●	●			
Hawk Inn & Mountain Resort	P.O. Box 64, Rte. 100, Plymouth, VT 05056; 802/672-3811, 1-800/685-4295 (U.S.)	●	●	●	●		●	●		●	●
The Homestead	Hot Springs, VA 24445; 703/839-5500, 1-800/542-5734 (VA), 1-800/336-5771 (U.S.)					●	●	●	●	●	●
Longboat Key Club & Resort	301 Gulf of Mexico Dr., Longboat Key, FL 34228; 813/383-8821, 1-800/282-0113 (FL), 1-800/237-8821 (U.S.)	●	●	●	●	●				●	●
Marriott at Sawgrass	1000 TPC Blvd., Ponte Vedra Beach, FL 32082; 904/285-7777, 1-800/457-4653 (FL), 1-800/228-9290 (U.S.)					●	●	●	●	●	●
PGA National Resort	400 Ave. of the Champions, Palm Beach Gardens, FL 33418; 407/627-2000, 1-800/633-9150	●	●	●	●	●				●	●
The Ritz-Carlton, Amelia Island	4750 Amelia Island Pkwy., Amelia Island, FL 32034; 904/277-1100, 1-800/241-3333	●	●	●	●	●				●	●
The Ritz-Carlton, Naples	280 Vanderbilt Beach Rd., Naples, FL 33963; 813/598-3300, 1-800/241-3333	●	●	●	●	●				●	●
The Ritz-Carlton, Palm Beach	100 S. Ocean Blvd., Manalapan, FL 33462; 407/533-6000, 1-800/241-3333	●	●	●	●	●				●	●
Turnberry Isle Resort & Club	19999 W. Country Club Dr., Aventura (N. Miami), FL 33180; 305/932-6200, 1-800/327-7028 (U.S.)	●	●	●	●	●	●			●	●
Westin Resort Hilton Head Island	2 Grasslawn Ave., Hilton Head Island, SC 29928; 803/681-4000, 1-800/933-3102	●	●	●	●	●	●			●	●
WESTERN U.S.											
The Arizona Biltmore	2400 E. Missouri Ave., Phoenix, AZ 85016; 602/955-6600, 1-800/950-0086					●	●	●	●		H
The Boulders Resort	34631 N. Tom Darlington Dr., Carefree, AZ 85377; 602/488-9009, 1-800/553-1717					●	●	●	●		●
The Broadmoor	P.O. Box 1439 (1 Lake Ave.), Colorado Springs, CO 80906; 719/634-7711, 1-800/634-7711	●				●	●	●	●	●	●
Flathead Lake Lodge	Flathead Lake Lodge Rd., Bigfork, MT 59911; 406/837-4391	●	●	●	●		●	●			●
Four Seasons Biltmore	1260 Channel Dr., Santa Barbara, CA 93108; 805/969-2261, 1-800/332-3442	●	●	●	●	●				●	●
Inn at Spanish Bay	2700 17 Mile Dr., Pebble Beach, CA 93953; 408/647-7500, 1-800/654-9300	●				●	●		●		
Jasper Park Lodge	P.O. Box 40 (Jasper Park Lodge Rd.), Jasper, Alberta T0E1E0; 403/852-3301, 1-800/528-0444 (Can. & U.S.)	●	●			●	●	●	●	●	H
La Quinta Hotel Golf & Tennis Resort	49499 Eisenhower Dr., P.O. Box 69, La Quinta, CA 92253; 619/564-4111, 1-800/472-4316 (CA), 1-800/854-1271 (U.S.)					●	●	●	●		H
The Lodge at Pebble Beach	17 Mile Dr., P.O. Box 1128, Pebble Beach, CA 93953; 408/624-3811, 1-800/654-9300	●	●	●	●		●			●	H
Loews Ventana Canyon Resort	7000 N. Resort Dr., Tucson, AZ 85715; 602/299-2020, 1-800/234-5117					●	●	●	●		●
Marriott's Camelback Inn Resort, Golf Club & Spa	5402 E. Lincoln Dr., Scottsdale, AZ 85253; 602/948-1700, 1-800/24-CAMEL					●	●	●	●		● ●
Quail Lodge Resort & Golf Club	8205 Valley Greens Dr., Carmel, CA 93923; 408/624-1581, 1-800/538-9516	●				●	●	●	●		●
Rancho Bernardo Inn	17550 Bernardo Oaks Dr., San Diego, CA 92128; 619/487-1611, 1-800/542-6096					●	●	●	●		● ●
The Ritz-Carlton, Laguna Niguel	33533 Ritz-Carlton Dr., Dana Point, CA 92629; 714/240-2000, 1-800/241-3333	●	●	●	●	●				●	●
Stein Eriksen Lodge	7700 Stein Way, Park City, UT 84060; 801/649-3700, 1-800/453-1302 (U.S.)					●	●	●	●		H
Tall Timber	S.S.R. Box 90, Durango, CO 81301; 303/259-4813						●	●		●	
The Wigwam Resort & Country Club	300 E. Indian School Lane, Litchfield Park, AZ 85340; 602/935-3811, 1-800/327-0396					●	●	●	●		● ●

H = holidays only

Theme Parks

Facilities
- Handicapped Access
- Picnic Facilities
- Strollers
- Baby Changing Facilities
- Locker Facilities

Tickets
- Senior
- Child
- Adult
- Season
- Multi-day
- One-day

State/Province	Park Name	Address/Telephone	Season (open daily)	Hours*	One-day	Multi-day	Season	Adult	Child	Senior	Locker Facilities	Baby Changing Facilities	Strollers	Picnic Facilities	Handicapped Access
California	Disneyland	1313 Harbor Blvd, Anaheim, CA 92803; 714/999-4000	Year-round	S 8AM-1AM W 10AM-varies	•	•	•	•	•	•	•	•	•	•	•
California	Great America	PO Box 1776 (95052), Great America Pkwy, Santa Clara, CA 95054; 408/988-1800	Mar.-Sept.	10AM-varies	•		•	•	•	•	•	•	•	•	•
California	Knott's Berry Farm	8039 Beach Blvd, Buena Park, CA 90620; 714/220-5200	Year-round	10AM-varies	•		•	•	•	•	•	•	•	•	•
California	Marine World Africa USA	Marine World Pkwy, Vallejo, CA 94589 707/643-ORCA, 644-4000	Year-round	9:30AM-varies	•	•	•	•	•	•	•	•	•	•	•
California	Sea World of California	1720 S Shores Rd, Sea World Dr & I-5, San Diego, CA 92109; 619/226-3901, 222-6363	Year-round	9AM-varies	•		•	•	•	•	•	•	•	•	•
California	Six Flags Magic Mountain	Box 5500, Magic Mtn. Pkwy & I-5, Valencia, CA 91385; 805/255-4100, 255-4111	Mem. Day-Labor Day	10AM-varies	•		•	•	•	•	•	•	•	•	•
Florida	Busch Gardens	PO Box 9158(33674), 3605 Bougainvillea, Tampa, FL 33674-9158; 813/987-5000	Year-round	varies	•		•	•	•	•	•	•	•	•	•
Florida	Sea World of Florida	7007 Sea World Dr, Orlando, FL 32821-8097; 407/351-3600	Year-round	S 8:30AM-10PM W 9AM-8PM	•	•	•	•	•	•	•	•	•	•	•
Florida	Universal Studios Florida	1000 Universal Studios Plaza, Orlando, FL 32819; 407/363-8000	Year-round	S 9AM-11PM W 9AM-7PM	•	•	•	•		•		•	•	•	•
Florida	Walt Disney World Resort	PO Box 10000, Lake Buena Vista, FL 32830-1000; 407/824-4321	Year-round	9AM-varies	•	•	•	•		•		•	•	•	•
Georgia	Six Flags Over Georgia	PO Box 43187, Atlanta, GA 30378; (7561 Six Flags Pkwy, Austell, 30001); 404/948-9290	Mar.-Nov.	10AM-varies	•		•	•	•	•	•	•	•	•	•
Illinois	Six Flags Great America	PO Box 1776, Grand Ave & I-94, Gurnee, IL 60031; 708/249-1776	May-Aug.	10AM-varies	•		•	•	•	•	•	•	•	•	•
Minnesota	Valleyfair	One Valleyfair Dr, Shakopee, MN 55379; 612/445-7600	May-Sept.	10AM-varies	•		•	•	•	•	•	•	•	•	•
Missouri	Worlds of Fun	4545 Worlds of Fun Ave, Kansas City, MO 64161; 816/454-4545	May-Sept.	10AM-varies	•		•	•	•	•	•	•	•	•	•
New Jersey	Six Flags Great Adventure	PO Box 120, NJ 537, Jackson, NJ 08527; 908/928-1821	Easter-Oct.	10AM-varies	•		•	•	•	•	•	•	•	•	•
New York	The Great Escape	PO Box 511, US 9, Lake George, NY 12845; 518/792-6568	Mem Day-Labor Day	9:30AM-6PM	•		•	•	•		•	•	•	•	•
North Carolina	Carowinds	PO Box 410289, Carowinds Blvd, Charlotte, NC 28241-0289; 800/822-4428, 704/588-2600	Mid Mar.-Mid Oct.	10AM-varies	•		•	•	•	•	•	•	•	•	•
Ohio	Cedar Point	PO Box 5006, Cedar Point, Sandusky, OH 44871-8006; 419/626-0830	May-Sept.	9AM-varies	•	•	•	•	•	•	•	•	•	•	•
Ohio	Kings Island	6300 Kings Island Dr, Kings Island, OH 45034; 513/398-5600	May-Sept.	9AM-varies	•	•	•	•	•	•	•	•	•	•	•
Ohio	Sea World of Ohio	1100 Sea World Dr, Aurora (Cleveland), OH 44202-8700; 800/63-SHAMU, 216/562-8101	May-Sept.	10AM-varies	•		•	•	•	•	•	•	•	•	•
Ontario	Boblo Island	3509 Biddle Ave., Wyandotte, MI 48219; 519/252-4444	Late May-Early Sept.	9AM-varies	•		•	•	•	•		•	•	•	•
Ontario	Canada's Wonderland	PO Box 624, Maple, ON L6A 1S6; 416/832-7000	Late May-Early Sept.	10AM-varies	•		•	•	•	•		•	•	•	•
Pennsylvania	Hersheypark	100 W Hersheypark Dr, Hershey, PA 17033; 800/HERSHEY	May-Sept.	10:30AM-varies	•		•	•	•	•	•	•	•	•	•
Tennessee	Dollywood	1020 Dollywood Lane, Pigeon Forge, TN 37865-4101; 615/428-9488	May-Oct.	9AM-varies	•		•	•	•	•	•	•	•	•	•
Tennessee	Opryland	2802 Opryland Dr, Nashville, TN 37214; 615/889-6611	May-Sept.	9AM-varies	•	•	•	•	•		•	•	•	•	•
Texas	AstroWorld	9001 Kirby Dr, Houston, TX 77054; 713/799-1234	Mem. Day-Labor Day	10AM-varies	•		•	•	•	•	•	•	•	•	•
Texas	Sea World of Texas	10500 Sea World Drive, San Antonio, TX 78251; 512/523-3611	Mar.-Nov.	10AM-varies	•	•	•	•	•	•	•	•	•	•	•
Texas	Six Flags Over Texas	PO Box 191, I-30 at TX 360, Arlington, TX 76010; 817/640-8900	May-Aug.	10AM-varies	•		•	•	•	•	•	•	•	•	•
Virginia	Busch Gardens/ Williamsburg	1 Busch Gardens Blvd, Williamsburg, VA 23187-8785; 804/253-3350	Apr.-Nov.	10AM-varies	•		•	•	•	•	•	•	•	•	•
Virginia	Kings Dominion	I-95 & VA 30, Doswell, VA 23047; 804/876-5000	Mar.-Oct.	9:30AM-varies	•		•	•	•	•	•	•	•	•	•

*Hours subject to change. Phone before you visit.   S Summer   W Winter

Fishing/Hunting

State	Fishing/Hunting Information	Resident Fishing	Resident Hunting	Resident Combination	Nonresident Fishing	Nonresident Hunting	Nonresident Combination Fishing/Hunting
Alabama	Alabama Game & Fish Division, 205/242-3466	$9.50	$16.00	$24.50	$16.00	$177.00	
Alaska	Alaska Dept. of Fish & Game, 907/465-4112	10.00	12.00	22.00	50.00	85.00	$135.00
Arizona	Arizona Game & Fish Dept., 602/942-3000	12.00	18.00	34.00	38.00	85.50	137.50
Arkansas	Arkansas Game & Fish Commission, 501/223-6300	10.50	10.50*	35.50	25.00	50.00*	
California	California Fish & Game, 916/739-3380	23.65	24.15		63.55	83.75	
Colorado	Colorado Division of Wildlife, 303/297-1192	20.25	15.25*	30.25	40.25	40.25*	
Connecticut	Connecticut Environmental Protection, 203/566-4409	15.00	10.00	21.00	25.00	42.00	55.00
Delaware	Delaware Fish & Wildlife, fishing: 302/739-3441; hunting 302/739-5297	8.50	12.50		15.00	45.00	
District of Columbia	Dept. of Consumer Regulatory Affairs, 202/404-1157	5.00	No hunting		7.50	No hunting	
Florida	Florida Game & Freshwater Fish Commission, 904/488-4676	12.00	11.00	22.00	30.00	150.00	
Georgia	Georgia Dept. of Natural Resources, 404/975-4230	9.00	10.00	18.00	24.00	59.00	
Hawaii	Hawaii Div. of Conservation & Resources Enforcement, 808/587-0077	3.75	10.00		7.50	95.00	
Idaho	Idaho Dept. of Fish & Game, 800/635-5355	16.00	7.00	21.00	41.00	86.00*	
Illinois	Illinois Dept. of Conservation, 217/782-2965	7.50	7.50	13.75	15.50	45.75	
Indiana	Indiana Division of Fish & Wildlife, 317/232-4080	8.75	8.75	13.75	15.75	40.75	
Iowa	Iowa Dept. of Natural Resources, 515/281-5145	10.50	12.50	23.50	22.50	60.50*	
Kansas	Kansas Dept. of Wildlife & Parks, 913/296-2281	10.50	10.50	21.50	25.50	50.50	
Kentucky	Kentucky Dept. of Fish & Wildlife, 502/564-4224 or -4762	8.50	8.50	15.00	20.00	75.00	
Louisiana	Louisiana Dept. of Wildlife & Fisheries, 800/442-2511	5.50	10.50		15.50	75.50	
Maine	Maine Dept. of Inland Fisheries & Wildlife, 207/289-5209	16.00	16.00*	29.00	43.00	48.00*	108.00
Maryland	Maryland Forest, Park & Wildlife Service, 301/974-3195	8.00	15.50	24.50	15.00	120.50**	
Massachusetts	Massachusetts Division Fish & Wildlife, 508/792-7270	17.50	17.50	24.50	22.50	28.50*	
Michigan	Michigan Dept. of Natural Resources, 517/373-1204	9.85	9.85*	45.35	20.35	50.35*	
Minnesota	Minnesota Dept. of Natural Resources, 612/296-6157	13.00	15.00*	29.00	28.50	61.00*	
Mississippi	Mississippi Dept. of Wildlife, Fisheries & Parks, 601/ 364-2120	4.00		13.00	20.00	150.00	
Missouri	Missouri Dept. of Conservation, 314/751-4115	8.00	8.00	14.00	25.00	50.00*	
Montana	Montana Dept. of Fish, Wildlife & Parks, 406/444-2535	11.00	**		45.00	**	
Nebraska	Nebraska Game & Parks Commission, 402/471-0641	11.50	8.50	19.50	25.00	40.00*	
Nevada	Nevada Dept. of Wildlife, 702/688-1500	15.50	20.50	34.00	45.50	100.50	
New Hampshire	New Hampshire Fish & Game, 603/271-3421	23.25	15.50	30.50	35.50	70.50	
New Jersey	New Jersey Division of Fish, Game & Wildlife, 609/292-2965	15.00	20.00	55.00	23.00	100.00	
New Mexico	New Mexico Dept. of Game & Fish, 505/827-7911	14.00	19.00	31.50	41.00	**	
New York	New York Dept. of Environmental Conservation, 518/457-5400	14.00	11.00*		28.00	41.00*	
North Carolina	North Carolina Wildlife Resources Commission, 919/662-4370	15.00	30.00	40.00	30.00	80.00	130.00
North Dakota	North Dakota State Game & Fish Dept., 701/221-6300	9.00	6.00*	25.00	20.00	56.00*	
Ohio	Ohio Division of Wildlife, 614/265-6300	12.00	12.00		19.00	81.00	
Oklahoma	Oklahoma Dept. of Wildlife Conservation, 405/521-4631	10.25	10.25	18.50	23.50	74.50*	
Oregon	Oregon Dept. of Fish & Wildlife, 503/229-5410	14.75	12.50	24.75	35.75	125.50	
Pennsylvania	Pennsylvania Game Commission, fishing: 717/657-4518; hunting: 717/787-4250	12.50	12.75		25.50	80.75	
Rhode Island	Rhode Island Division of Fish & Wildlife, 401/ 789-3094	9.50	9.50*	15.50	20.50	20.50*	
South Carolina	South Carolina Wildlife & Marine Resources, 803/734-3838	10.00	12.00*	17.00	34.00	75.00*	
South Dakota	South Dakota Game, Fish & Parks Dept., 605/773-3485	14.00	21.00	30.00	35.00	65.00	
Tennessee	Tennessee Wildlife Resources Agency, 615/781-6580			16.50	26.00	156.00*	
Texas	Texas Parks & Wildlife Dept., 512/389-4570	13.00	13.00	25.00	20.00	200.00	
Utah	Utah State Division of Wildlife Resources, 801/596-8660	18.00	12.00*	35.00	40.00	40.00*	
Vermont	Vermont Fish & Wildlife Dept., 802/244-7331	18.00	12.00	26.00	35.00	75.00	95.00
Virginia	Virginia Dept. of Game & Inland Fisheries, 804/367-1000	12.00	12.00		30.00	60.00*	
Washington	Washington Dept. of Wildlife, 206/753-5700	17.00	15.00	29.00	48.00	150.00	
West Virginia	West Virginia Dept. of Natural Resources, 304/348-2771	11.00	11.00	25.00	25.00	70.00	
Wisconsin	Wisconsin Dept. of Natural Resources, 608/266-1877	12.00	12.00*	38.00	28.00	70.00*	
Wyoming	Wyoming Game & Fish Dept., 307/777-4600	9.00	**		50.00	**	

©1993 Rand McNally & Company   *Annual fee for small game and/or birds.   **License fee depends on type of game.

Special Events

State/Province	Event	Information	Date(s)	Agricultural Fair/Livestock Show	Historical	Food	Parade	Arts & Crafts	Entertainment/Music	Sports
Alabama	City Stages	City Stages, P.O. Box 2266, Birmingham, AL 35201; 205/251-1272	Mid-June			•	•		•	
Alaska	Golden Days	Golden Days, Fairbanks C of C, 709 Second Ave., Fairbanks, AK 99701; 907/452-1105	Mid-July	•				•		•
Alberta	Klondike Days '93	Klondike Days Assn. 1660, 10020-101A Ave., Edmonton, AB T5J 3G2; 403/426-4055	Late July		•				•	•
Arizona	Navajo Nation Tribal Fair	Navajo Nation Fair Office, P.O. Box Drawer U, Window Rock, AZ 86515; 602/871-6702	Early Sept.	•	•	•			•	
Arkansas	Petit Jean Show '93	Petit Jean Show '93, c/o Museum of Automobiles, Route 3, Box 306, Morrilton, AR 72110; 501/727-5427	Mid-June						•	
British Columbia	Classic Boat Festival	Victoria Real Estate Board, 3035 Nanaimo St., Victoria, BC V8T 4W2; 604/385-7766	Early Sept.			•	•		•	•
California	Super Bowl XXVII	LA Super Bowl Host Committee, 515 S. Figueroa St., 11th Fl. Los Angeles, CA 90071; 213/663-1993	Mid-Jan.	•	•			•		
Colorado	Colorado Shakespeare Festival	Shakespeare Festival, University of Colorado at Boulder, Campus Box 261, Boulder, CO 80309-0261; 303/492-2782	Late June–Mid-Aug.		•				•	
Connecticut	New London Sail Festival	Marine Commerce & Development, 111 Union St., New London, CT 06320; 203/443-8331	Early July	•	•					•
Delaware	Old Dover Days	Old Dover Days, Friends of Old Dover, P.O. Box 44, Dover, DE 19903; 302/734-5250	Early May		•	•	•	•		
District of Columbia	Pageant of Peace	National Park Service, 1100 Ohio Dr., SW, Washington, DC 20242; 202/619-7222	Mid-Dec.		•				•	
Florida	Annual Maritime Festival	St. Augustine C of C, P.O. Drawer O, St. Augustine, FL 32085; 904/829-5681	Early Oct.	•			•		•	
Georgia	Antebellum Jubilee	Antebellum Jubilee, c/o Stone Mountain Park, P.O. Box 778, Stone Mountain, GA 30086; 404/498-5702	Late Mar.–Early Apr.		•	•			•	•
Hawaii	Aloha Festivals	Aloha Festivals, 750 Amana St., Suite 111, Honolulu, HI 96814; 808/944-8857	Mid-Sept.–Mid-Oct.	•	•			•		
Idaho	Three Island Crossing	Three Island State Park, P.O. Box 609, Glenns Ferry, ID 83623; 208/366-2394	Early Aug.		•			•	•	•
Illinois	Great River Ramble	Quad City C&V Bureau, P.O. Box 3546, Rock Island, IL 61201; 800/747-7800 or 309/788-7800	Late June–Early July	•	•			•		
Indiana	Covered Bridge Festival	Parke County C&V Bureau, P.O. Box 165, Rockville, IN 47872; 317/569-5226	Mid-Oct.			•	•		•	•
Iowa	Nordic Fest	Nordic Fest, Inc., P.O. Box 364, Decorah, IA 52101; 319/382-3990	Late July			•			•	
Kansas	Smoky Hill River Festival	Festival Office, P.O. Box 2181, Salina, KS 67402; 913/826-7410	Mid-June	•	•	•		•		
Kentucky	Maifest '93	MainStrasse Village Assn., 616 Main Strasse, Covington, KY 41011; 606/491-0458	Mid-May	•	•	•		•		
Louisiana	New Orleans Jazz & Heritage Festival	Jazz Fest, 1205 N. Rampart St., New Orleans, LA 70116; 504/522-4786	Late April–Early May		•	•		•		
Maine	Fryeburg Fair	West Oxford Agricultural Society, Box 36, Fryeburg, ME 04037; 207/935-3268	Early Oct.							•
Manitoba	Folklorama	Folklorama, 375 York Ave., Winnipeg, MB R3C 3J3; 800/665-0234 (U.S.) 204/944-9793	Early Aug.	•	•	•		•		
Maryland	Autumn Glory Festival	Garrett County Promotions Office, Court House, 200 S. Third St., Oakland, MD 21550; 301/334-1948	Mid-Oct.			•	•	•		
Massachusetts	Boston Harborfest	Boston Harborfest, 45 School St., Boston, MA 02116; 617/227-1528	Early July		•				•	•
Michigan	Tulip Time Festival	Tulip Time Festival, Inc., 171 Lincoln Ave., Holland, MI 49423; 616/396-4221	Early May		•		•		•	
Minnesota	Festival of Nations	Festival of Nations, 1694 Como Ave., St. Paul, MN 55108; 612/647-0191	Early May		•	•				
Mississippi	Jubilee Jam	Arts Alliance of Jackson-Hinds County, P.O. Box 17, Jackson, MS 39205; 601/960-1557	Mid-May		•			•		
Missouri	Maifest	Maifest Council, Box 51, Hermann, MO 65041; 314/486-2744	Mid-May		•	•	•	•	•	
Montana	Custer's Last Stand Re-enactment	Hardin C of C, 200 N. Center Ave., Hardin, MT 59034; 406/665-1672	Late June						•	

			Type of Celebration		Agricultural Fair/Livestock Show	Historical	Food	Parade	Arts & Crafts	Entertainment/Music	Sports
State/Province	**Event**	**Information**		**Date(s)**							
Nebraska	Czech Festival	Irma Ourecky, P.O. Box 652, Wilber, NE 68465; 402/821-2485		Late July–Early Aug.		•			•	•	•
Nevada	Great Reno Balloon Race	Great Reno Balloon Race, P.O. Box 12695, Reno, NV 89510; 702/826-1181		Mid-Sept.	•	•	•			•	
New Brunswick	Festival by the Sea	Festival by the Sea, P.O. Box 6848, Station A, Saint John, NB E2L 4S3; 506/632-0086		Mid-Aug.		•					
Newfoundland	Stephenville Festival	Stephenville Festival, P.O. Box 282, Stephenville, NF A2N 2Z4; 709/643-4982		Early July–Early Aug.		•					
New Hampshire	Annual League of N.H. Craftsmen Fair	League of N.H. Craftsmen, 205 N. Main St., Concord, NH 03301; 603/224-1471		Mid-Aug.		•	•				
New Jersey	Chatsworth Cranberry Festival	Festival Committee, 64 Mill St., Vincentown, NJ 08088; 609/859-9701		Mid-Oct.		•	•			•	•
New Mexico	Hatch Chile Festival	Hatch C of C, P.O. Box 38, Hatch, NM 87937; 505/267-5050		Early Sept.	•	•					
New York	World University Games	World University Games, 235 North St., Buffalo, NY 14201; 716/888-9300		Mid-July	•						
North Carolina	The Lost Colony Outdoor Drama	Lost Colony Drama, P.O. Box 40, Manteo, NC 27954; 800/488-5012 or 919/473-3414		Mid-June–Late Aug.		•				•	
North Dakota	Riverfront Days	Riverkeepers, 115 Roberts St., P.O. Box 171, Fargo, ND 58107; 701/235-2895		Mid-June	•		•	•			
Nova Scotia	Annapolis Valley Apple Blossom Festival	Dorothy Butt, Old Courthouse Museum, Kentville, NS B3N 2E2; 902/678-8322		Late May		•		•	•		
Ohio	Ohio River Sternwheel Festival	Marietta Tourist & Convention Bureau, 316 Third St., Marietta, OH 45750; 614/373-5178		Early Sept.	•	•					
Oklahoma	Cherokee Strip Centennial	Cherokee Strip Centennial Foundation, 401 E. Oklahoma, Enid, OK 73701; 405/233-4353		Mid-April Mid-Sept.	•		•			•	•
Ontario	Winterlude/Bal de Neige	National Capital Commission, 161 Laurier Ave., W, Ottawa, ON K1P 6J6; 800/267-0450 or 613/239-5145		Mid-Feb.	•	•	•				
Oregon	Peter Britt Gardens Music & Arts Festival	Peter Britt Festivals, P.O. Box 1124, Medford, OR 97501; 800/88-BRITT or 503/779-0847		Mid-June–Early Sept.		•	•				
Pennsylvania	Gettysburg Civil War Heritage Days	Gettysburg Travel Council, 35 Carlisle St., Gettysburg, PA 17325; 717/334-6274		Late June–Early July		•				•	
Prince Edward Island	Summerside Lobster Carnival	Summerside CofC, 263 Harbour Dr., Suite 10, Summerside, PE C1N 5P1; 902/436-9651		Early July	•	•	•	•	•		•
Québec	Montréal Intl. Jazz Festival	Montréal Intl. Jazz Festival, 355 St. Catherine St., W, Office 301, Montréal, PQ H3B 1A5; 514/289-9472		Late June–Early July		•					
Rhode Island	JVC Jazz Festival-Newport	Jazz Festival, P.O. Box 605, Newport, RI 02840; 401/847-3700		Mid-Aug.		•					
Saskatchewan	Saskatoon Exhibition	Ed Sikorski, Prairieland Exhibition Corp., Box 6010, Saskatoon, SK F7K 4E4; 306/931-7149		Mid-July	•	•		•	•		•
South Carolina	Freedom Weekend Aloft	Keri Hall, 135 S. Main St., Suite LL1, Greenville, SC 29601; 803/232-3700		Late June–Early July	•	•	•				
South Dakota	Days of '76	Days of '76 Committee, 735 Main St., Deadwood, SD 57732; 605/578-1876		Early Aug.	•			•		•	•
Tennessee	Tennessee Valley Fair	Fair Office, P.O. Box 6066, Knoxville, TN 37914; 615/637-5840		Mid-Sept.	•				•		•
Texas	George Washington's Birthday Celebration	Jody Powell, P.O. Box 816, Laredo, TX 78042; 512/722-0589		Mid-Feb.		•		•	•	•	
Utah	Christmas Lights at Temple Square	Temple Square Public Relations, 50 W. North Temple, Salt Lake City, UT 84150; 801/240-4869		Late Nov.–Late Dec.							
Vermont	Northeast Kingdom Annual Fall Foliage Festival	Send SASE to: Fall Festival Committee, Box 38, West Danville, VT 05873; 802/563-2472		Late Sept.			•			•	
Virginia	Harborfest '93	Harborfest '93, 120 W. Main St., Norfolk, VA 23510; 804/627-5329		Early June	•	•		•	•		
Washington	National Lentil Festival	Festival Committee, P.O. Box 424, Pullman, WA 99163; 800/ENJOY-IT		Mid-Sept.	•	•	•		•		
West Virginia	Winter Festival of Lights	Wheeling C&V Bureau, 1000 Boury Center, Wheeling, WV 26003; 800/828-3097 or 304/233-7709		Early Nov.–Late Feb.			•				
Wisconsin	Summerfest	Summerfest, 200 N. Harbor Dr., Milwaukee, WI 53202; 800/837-FEST		Late June–Early July		•	•		•		
Wyoming	Oregon Trail 150th Anniversary	Wyoming Div. of Tourism, I-25 at College Dr., Cheyenne, WY 82002-0240; 307/777-7777 or 800/225-5996		All year	•	•	•	•	•	•	•

C&V: Convention & Visitors (Bureau)
CofC: Chamber of Commerce

National Park Areas

		Accommodations			Facilities					Activities							
State/Province	**Park Name**	Campgrounds	Hotel, Motel, Lodge	Cabins	Handicap Access Rest Rooms	Restaurant, Snacks	Museum/Exhibit	Groceries, Ice	Bathhouses	Cross-Country Ski Trail	Snowmobile Route	Fishing	Boating	Swimming	Horseback Riding	Hiking	NPS Guided Tours
Alabama	Russell Cave Natl. Monument, Rte. 1, Box 175, Bridgeport, AL 35740		•	•										•		•	
	Tuskegee Institute Natl. Historic Site, P.O. Drawer 10, Tuskegee Institute, AL 36088	•												•		•	
Alaska	Denali Natl. Park and Preserve, P.O. Box 9, McKinley Park, AK 99755	•	•				•			•			•	•		•	•
	Glacier Bay Natl. Park and Preserve, P.O. Box 140, Gustavus, AK 99826	•	•			•	•						•			•	•
	Katmai Natl. Park and Preserve, P.O. Box 7, King Salmon, AK 99613	•	•			•	•			•			•			•	•
	Kenai Fjords Natl. Park, P.O. Box 1727, Seward, AK 99664	•	•			•	•	•	•								
	Lake Clark Natl. Park and Preserve, 4230 University Dr., Suite 311, Anchorage, AK 99508		•			•	•	•	•				•		•	•	
	Sitka Natl. Historical Park, P.O. Box 738, Sitka, AK 99835	•	•				•					•					
	Wrangell-St. Elias Natl. Park and Preserve, P.O. Box 29, Glennallen, AK 99588		•	•			•									•	
Alberta	Banff Natl. Park, Box 900, Banff, AB T0L 1E0	•	•	•	•	•	•		•	•	•	•	•	•	•	•	•
	Jasper Natl. Park, Box 10, Jasper, AB T0E 1E0	•	•	•	•	•	•		•	•	•	•	•	•	•	•	•
	Waterton Lakes Natl. Park, Waterton Park, AB T0K 2M0	•	•	•	•	•	•		•		•	•	•	•	•	•	•
Arizona	Canyon de Chelly Natl. Monument, P.O. Box 588, Chinle, AZ 86503	•	•	•			•				•	•			•	•	•
	Casa Grande Natl. Monument, P.O. Box 518, Coolidge, AZ 85228	•					•							•			
	Chiricahua Natl. Monument, Dos Cabezas Route, Box 6500, Willcox, AZ 85643	•	•				•							•			•
	Grand Canyon Natl. Park, P.O. Box 129, Grand Canyon, AZ 86023	•	•	•			•				•	•	•	•	•	•	•
	Hubbell Trading Post Natl. Historic Site, P.O. Box 150, Ganado, AZ 86505	•					•							•			
	Montezuma Castle Natl. Monument, P.O. Box 219, Camp Verde, AZ 86322						•							•			
	Navajo Natl. Monument, H.C. 71, Box 3, Tonalea, AZ 86044-9704	•		•			•							•			•
	Organ Pipe Cactus Natl. Monument, Rte. 1, Box 100, Ajo, AZ 85321	•	•				•							•			
	Petrified Forest Natl. Park, Petrified Forest Natl. Park, AZ 86028		•				•					•		•			
	Pipe Spring Natl. Monument, Moccasin, AZ 86022	•					•					•		•			
	Saguaro Natl. Monument, 36933 Old Spanish Trail, Tucson, AZ 85730	•	•				•							•			
	Tonto Natl. Monument, P.O. Box 707, Roosevelt, AZ 85545	•					•							•			
	Tumacacori Natl. Historical Park, P.O. Box 67, Tumacacori, AZ 85640	•					•							•			
	Tuzigoot Natl. Monument, P.O. Box 68, Clarkdale, AZ 86324						•							•			
	Walnut Canyon Natl. Monument, Walnut Canyon Rd., Flagstaff, AZ 86004-9705						•							•			
	Wupatki Natl. Monument, H.C. 33, Box 444A, Flagstaff, AZ 86004	•	•				•							•			
Arkansas	Fort Smith Natl. Historic Site, P.O. Box 1406, Fort Smith, AR 72902	•					•							•			
	Hot Springs Natl. Park, P.O. Box 1860, Hot Springs, AR 71902	•	•	•			•		•			•		•		•	•
British Columbia	Mount Revelstoke/Glacier Natl. Park, Box 350, Revelstoke, BC V0E 2S0	•	•				•	•				•	•	•		•	•
California	Cabrillo Natl. Monument, P.O. Box 6670, San Diego, CA 92106	•	•				•					•					
	Channel Islands Natl. Park, 1901 Spinnaker Dr., Ventura, CA 93001	•	•		•	•	•					•		•			•
	Death Valley Natl. Monument (Calif., Nev.), Death Valley, CA 92328	•	•	•	•		•				•	•	•	•	•	•	•
	Devils Postpile Natl. Mon., c/o Sequoia and Kings Canyon Natl. Parks, Three Rivers, CA 93271	•	•	•	•		•					•				•	
	Fort Point Natl. Historic Site, P.O. Box 29333, Presidio of San Francisco, CA 94129	•					•					•		•			
	Golden Gate Natl. Recreation Area, Fort Mason, Bldg. 201, San Francisco, CA 94123	•	•	•	•		•			•		•		•		•	•
	Joshua Tree Natl. Monument, 74485 National Monument Dr., Twentynine Palms, CA 92277	•	•	•			•					•		•		•	•
	Kings Canyon Natl. Park, Three Rivers, CA 93271	•	•	•			•		•		•	•	•	•	•	•	•
	Lassen Volcanic Natl. Park, Mineral, CA 96063	•	•	•	•	•	•		•		•	•	•	•	•	•	
	Lava Beds Natl. Monument, P.O. Box 867, Tulelake, CA 96134	•	•				•					•		•			•
	Muir Woods Natl. Monument, Mill Valley, CA 94941	•	•				•							•	•		
	Pinnacles Natl. Monument, Paicines, CA 95043		•				•					•		•			
	Redwood Natl. Park, 1111 2nd St., Crescent City, CA 95531	•	•	•	•		•					•		•		•	•
	Santa Monica Mts. Natl. Recreation Area, 30401 Agoura Rd., Suite 100, Agoura Hills, CA 91301	•	•	•	•	•	•				•	•	•	•		•	•
	Sequoia Natl. Park, Three Rivers, CA 93271	•	•	•			•		•		•	•	•	•	•	•	•
	Whiskeytown-Shasta-Trinity Natl. Recreation Area, P.O. Box 188, Whiskeytown, CA 96095	•	•	•	•	•	•		•		•	•		•		•	•
	Yosemite Natl. Park, P.O. Box 577, Yosemite Natl. Park, CA 95389	•	•	•	•	•	•	•		•	•	•	•	•	•	•	•
Colorado	Black Canyon of the Gunnison Natl. Monument, P.O. Box 1648, Montrose, CO 81402	•	•				•	•				•		•		•	•
	Colorado Natl. Monument, Fruita, CO 81521	•	•	•			•							•		•	•
	Curecanti Natl. Recreation Area, 102 Elk Creek, Gunnison, CO 81230	•	•		•	•	•	•	•		•	•		•		•	•
	Dinosaur Natl. Monument (Colo., Utah), P.O. Box 210, Dinosaur, CO 81610	•	•			•	•				•		•		•	•	

Activities, Facilities, and Accommodations in Selected National Park Service areas of the United States and Canada.

Column groupings (read top-to-bottom in the original header):

- **Accommodations:** Campgrounds; Hotel, Motel, Lodge; Cabins
- **Facilities:** Handicap Access Rest Rooms; Restaurant, Snacks; Museum/Exhibit; Groceries, Ice; Bathhouses
- **Activities:** Cross-Country Ski Trail; Snowmobile Route; Fishing; Boating; Swimming; Horseback Riding; Hiking; NPS Guided Tours

State/Province	Park Name	Campgrounds	Hotel, Motel, Lodge	Cabins	Handicap Access Rest Rooms	Restaurant, Snacks	Museum/Exhibit	Groceries, Ice	Bathhouses	Cross-Country Ski Trail	Snowmobile Route	Fishing	Boating	Swimming	Horseback Riding	Hiking	NPS Guided Tours
	Florissant Fossil Beds Natl. Monument, P.O. Box 185, Florissant, CO 80816				•		•					•				•	•
	Great Sand Dunes Natl. Monument, Mosca, CO 81146	•			•		•	•						•		•	•
	Mesa Verde Natl. Park, Mesa Verde Natl. Park, CO 81330	•	•		•	•	•	•									•
	Rocky Mountain Natl. Park, Estes Park, CO 80517	•			•	•	•			•	•	•			•	•	•
District of Columbia	Ford's Theatre Natl. Historic Site, c/o NCP-Central, 900 Ohio Dr., SW, Washington, DC 20242						•										•
	John F. Kennedy Center for the Performing Arts, Natl. Park Service, 2700 F. St., NW, Washington, DC 20566					•	•										•
	Lincoln Memorial, c/o NCP-Central, 900 Ohio Dr., SW, Washington, DC 20242																•
	National Mall, c/o Natl. Capital Region, 1100 Ohio Dr., SW, Washington, DC 20242					•	•					•					•
	Thomas Jefferson Memorial, c/o NCP-Central, 900 Ohio Dr., SW, Washington, DC 20242						•										•
	Washington Monument, c/o NCP-Central, 900 Ohio Dr., SW, Washington, DC 20242					•	•										•
	White House, c/o NCR, Natl. Park Service, 1100 Ohio Dr., SW, Washington, DC 20242						•										•
Florida	Biscayne Natl. Park, P.O. Box 1369, Homestead, FL 33090	•			•		•					•	•	•		•	•
	Canaveral Natl. Seashore, 2532 Garden St., Titusville, FL 32796				•		•					•	•	•		•	•
	Castillo de San Marcos Natl. Monument, 1 Castillo Dr., St. Augustine, FL 32084				•		•										•
	Everglades Natl. Park, P.O. Box 279, Homestead, FL 33030	•	•	•	•	•	•	•	•			•	•			•	•
	Fort Jefferson Natl. Monument, c/o Everglades Natl. Park, P.O. Box 279, Homestead, FL 33030	•					•					•	•	•			
	Fort Matanzas Natl. Monument, c/o Castillo de San Marcos Natl. Monument, 1 Castillo Dr., St. Augustine, FL 32084						•						•				•
	Gulf Islands Natl. Seashore, 1801 Gulf Breeze Pkwy., Gulf Breeze, FL 32561 (Also in Miss.)	•			•		•	•	•			•	•	•		•	•
Georgia	Andersonville Natl. Historic Site, Rte. 1, Box 85, Andersonville, GA 31711				•		•										•
	Chattahoochee River Natl. Recreation Area, 1978 Island Ford Pkwy., Dunwoody, GA 30350				•		•					•	•			•	•
	Cumberland Island Natl. Seashore, P.O. Box 806, St. Marys, GA 31558	•					•					•	•	•		•	•
	Fort Frederica Natl. Monument, Rte. 9, Box 286 C, St. Simons Island, GA 31522				•		•										•
	Fort Pulaski Natl. Monument, P.O. Box 30757, Savannah, GA 31410				•		•					•	•			•	•
	Martin Luther King, Jr., Natl. Historic Site, 522 Auburn Ave., NE, Atlanta, GA 30312						•										•
	Ocmulgee Natl. Monument, 1207 Emery Hwy., Macon, GA 31201				•		•					•				•	•
Hawaii	Haleakala Natl. Park, P.O. Box 369, Makawao, Maui, HI 96768	•		•			•								•	•	
	Hawaii Volcanoes Natl. Park, Hawaii Natl. Park, HI 96718	•	•	•	•	•	•	•								•	
	USS Arizona Memorial, 1 Arizona Memorial Place, Honolulu, HI 96818				•		•										•
Idaho	Craters of the Moon Natl. Monument, P.O. Box 29, Arco, ID 83213	•					•									•	•
Illinois	Lincoln Home Natl. Historic Site, 413 S. Eighth St., Springfield, IL 62701						•										•
Indiana	Indiana Dunes Natl. Lakeshore, 1100 N. Mineral Springs Rd., Porter, IN 46304	•			•	•	•			•		•	•	•		•	•
Iowa	Effigy Mounds Natl. Monument, R.R. 1, Box 25A, Harpers Ferry, IA 52146						•					•				•	•
	Herbert Hoover Natl. Historic Site, P.O. Box 607, West Branch, IA 52358						•					•				•	•
Kansas	Fort Larned Natl. Historic Site, Rte. 3, Larned, KS 67550				•		•										•
	Fort Scott Natl. Historic Site, Old Fort Blvd., Fort Scott, KS 66701						•										•
Kentucky	Abraham Lincoln Birthplace Natl. Historic Site, 2995 Lincoln Farm Rd., Hodgenville, KY 42748				•		•									•	
	Mammoth Cave Natl. Park, Mammoth Cave, KY 42259	•	•	•	•	•	•	•				•	•		•	•	•
Louisiana	Jean Lafitte Natl. Historical Park and Preserve, 423 Canal St., New Orleans, LA 70130						•					•	•			•	•
Maine	Acadia Natl. Park, P.O. Box 177, Bar Harbor, ME 04609	•	•		•	•	•	•	•	•		•	•	•	•	•	•
Manitoba	Riding Mountain Natl. Park, Wasagaming, MB R0J 2H0	•	•	•	•	•	•	•	•	•		•	•	•	•	•	•
Maryland	Assateague Island Natl. Seashore (Md., Va.), Rte. 2, Box 294, Berlin, MD 21811	•							•			•	•	•		•	•
	Chesapeake & Ohio Canal Natl. Hist. Park (W.Va., Md., Va.), P.O. Box 4, Sharpsburg, MD 21782	•				•	•					•	•			•	•
	Fort McHenry Natl. Mon. and Historic Shrine, end of E. Fort Avenue, Baltimore, MD 21230-5393				•		•										•
Massachusetts	Adams Natl. Historic Site, 135 Adams St., P.O. Box 531, Quincy, MA 02269-0531																•
	Boston Natl. Historical Park, Charlestown Navy Yard, Boston, MA 02129					•	•										•
	Cape Cod Natl. Seashore, South Wellfleet, MA 02663				•		•			•		•	•	•		•	•
	John Fitzgerald Kennedy Natl. Historic Site, 83 Beals St., Brookline, MA 02146																•
	Salem Maritime Natl. Historic Site, Custom House, 174 Derby St., Salem, MA 01970						•					•				•	•
Michigan	Isle Royale Natl. Park, 87 N. Ripley St., Houghton, MI 49931	•	•	•			•					•	•			•	•
	Pictured Rocks Natl. Lakeshore, P.O. Box 40, Munising, MI 49862	•					•			•		•	•	•		•	

Activities, Facilities, and Accommodations in Selected National Park Service areas of the United States and Canada.

State/Province	Park Name	Campgrounds	Hotel, Motel, Lodge	Cabins	Handicap Access Rest Rooms	Restaurant, Snacks	Museum/Exhibit	Groceries, Ice	Bathhouses	Cross-Country Ski Trail	Snowmobile Route	Fishing	Boating	Swimming	Horseback Riding	Hiking	NPS Guided Tours
	Sleeping Bear Dunes Natl. Lakeshore, 9922 Front St., P.O. Box 277, Empire, MI 49630	•	•		•	•	•		•	•		•	•				•
Minnesota	Grand Portage Natl. Monument, P.O. Box 666, Grand Marais, MN 55604	•	•				•		•			•		•			
	Pipestone Natl. Monument, P.O. Box 727, Pipestone, MN 56164						•					•		•			
	Voyageurs Natl. Park, HCR 9, Box 600, International Falls, MN 56649	•	•				•		•	•		•	•	•	•		•
Mississippi	Natchez Trace Parkway (Miss., Ala., Tenn.), R.R. 1, NT-143, Tupelo, MS 38801		•		•	•	•		•			•		•			
Missouri	George Washington Carver Natl. Monument, P.O. Box 38, Diamond, MO 64840	•					•					•		•			
	Harry S Truman Natl. Historic Site, 223 N. Main St., Independence, MO 64050	•					•					•		•			
	Jefferson Natl. Expansion Memorial, 11 North 4th St., St. Louis, MO 63102	•					•					•		•			
	Ozark Natl. Scenic Riverways, P.O. Box 490, Van Buren, MO 63965	•	•	•	•	•	•	•		•		•	•	•	•		•
Montana	Bighorn Canyon Natl. Recreation Area (Mont., Wyo.), P.O. Box 458, Fort Smith, MT 59035	•	•		•	•	•	•	•			•	•	•	•		•
	Little Big Horn Battlefield Natl. Monument, P.O. Box 39, Crow Agency, MT 59022	•					•					•		•			
	Glacier Natl. Park, West Glacier, MT 59936	•	•	•	•	•	•	•		•		•	•	•	•	•	•
Nebraska	Agate Fossil Beds Natl. Monument, P.O. Box 27, Gering, NE 69341		•				•					•		•			
Nevada	Great Basin Natl. Park, Baker, NV 89311	•	•	•			•		•			•		•			•
	Lake Mead Natl. Recreation Area (Nev., Ariz.), 601 Nevada Hwy., Boulder City, NV 89005-2426		•		•	•	•	•		•	•	•	•	•		•	•
New Hampshire	Saint-Gaudens Natl. Historic Site, R.R. 3, Box 73, Cornish, NH 03745-9704	•	•				•					•		•			
New Jersey	Edison Natl. Historic Site, Main St. and Lakeside Ave., West Orange, NJ 07052	•										•		•			
New Mexico	Aztec Ruins Natl. Monument, P.O. Box 640, Aztec, NM 87410						•										
	Bandelier Natl. Monument, HCR 1, Box 1, Suite 15, Los Alamos, NM 87544	•	•				•			•	•	•	•				•
	Capulin Volcano Natl. Monument, Capulin, NM 88414		•									•		•			
	Carlsbad Caverns Natl. Park, 3225 National Parks Hwy., Carlsbad, NM 88220	•	•									•	•	•			
	Chaco Culture Natl. Historical Park, Star Route 4, Box 6500, Bloomfield, NM 87413	•	•									•		•			•
	El Morro Natl. Monument, Rte. 2, Box 43, Ramah, NM 87321-9603		•									•		•			
	Gila Cliff Dwellings Natl. Monument, Rte. 11, Box 100, Silver City, NM 88061							•				•		•			
	Pecos Natl. Historical Park, P.O. Drawer 418, Pecos, NM 87522	•										•		•			
	Salinas Pueblo Missions Natl. Monument, P.O. Box 496, Mountainair, NM 87036	•										•		•			
	White Sands Natl. Monument, P.O. Box 458, Alamogordo, NM 88310	•	•									•		•			
New York	Castle Clinton Natl. Monument, c/o Manhattan Sites, NPS, 26 Wall St. New York, NY 10005	•										•		•			
	Fire Island Natl. Seashore, 120 Laurel St., Patchogue, NY 11772	•	•		•	•	•		•	•		•	•	•			
	Fort Stanwix Natl. Monument, 112 E. Park St., Rome, NY 13440	•										•		•			
	Gateway Natl. Recreation Area (N.Y., N.J.), Floyd Bennett Field, Bldg. 69, Brooklyn, NY 11234	•	•	•	•	•	•		•		•	•	•	•			
	Home of Franklin D. Roosevelt Natl. Historic Site, 519 Albany Post Rd., Hyde Park, NY 12538						•					•		•			
	Sagamore Hill Natl. Historic Site, 20 Sagamore Hill Rd., Oyster Bay, NY 11771						•					•		•			
	Statue of Liberty Natl. Monument (N.Y., N.J.), Liberty Island, New York, NY 10004	•										•	•	•			
North Carolina	Blue Ridge Parkway (N.C., Va.), 200 BB&T Bldg., One Pack Square, Asheville, NC 28801	•	•	•		•	•	•	•			•	•		•	•	•
	Cape Hatteras Natl. Seashore, Rte. 1, Box 675, Manteo, NC 27954	•	•		•	•	•		•	•	•	•	•	•			
	Cape Lookout Natl. Seashore, 3601 Bridges St., Suite F, Morehead City, NC 28557	•	•		•	•	•					•			•		
	Carl Sandburg Home Natl. Historic Site, 1928 Little River Rd., Flat Rock, NC 28731	•	•									•		•			
	Fort Raleigh Natl. Historic Site, Cape Hatteras Group, Rte. 1, Box 675, Manteo, NC 27954	•										•		•			
North Dakota	Fort Union Trading Post Natl. Historic Site (N. Dak., Mont.), Buford Route, Williston, ND 58801	•	•			•	•	•				•		•			
	Theodore Roosevelt Natl. Park, P.O. Box 7, Medora, ND 58645	•	•	•		•	•	•				•		•			•
Nova Scotia	Fortress of Louisbourg Natl. Historic Park, Box 160, Louisbourg, NS B0A, 1M0	•	•						•			•	•	•			
Ohio	Cuyahoga Valley Natl. Recreation Area, 15610 Vaughn Rd., Brecksville, OH 44141	•	•	•	•		•		•			•		•		•	
	Mound City Group Natl. Monument, 16062 State Route 104, Chillicothe, OH 45601	•	•				•					•		•			
Oklahoma	Chickasaw Natl. Recreation Area, P.O. Box 201, Sulphur, OK 73086	•	•		•	•	•					•		•			•
Ontario	Georgian Bay Islands Natl. Park, P.O. Box 28, Honey Harbour, ON P0E 1E0	•	•			•	•		•			•		•			•
	St. Lawrence Islands Natl. Park, P.O. Box 469, RR3, Mallorytown Landing, ON K0E 1R0	•	•			•	•		•			•		•			
Oregon	Crater Lake Natl. Park, P.O. Box 7, Crater Lake, OR 97604	•	•				•	•	•	•		•	•	•	•	•	•
	Oregon Caves Natl. Monument, 19000 Caves Hwy., Cave Junction, OR 97523		•										•	•			
Pennsylvania	Delaware Water Gap Natl. Recreation Area (Pa., N.J.), Bushkill, PA 18324	•	•			•	•	•				•		•			•
	Eisenhower National Historic Site, Gettysburg, PA 17325	•										•		•			
	Gettysburg Natl. Military Park, Gettysburg, PA 17325	•	•									•		•			
	Hopewell Furnace Natl. Historic Site, 2 Mark Bird Lane, Box 345, Elverson, PA 19520		•									•	•	•			

Activities, Facilities, and Accommodations in Selected National Park Service areas of the United States and Canada.

		Activities								Facilities					Accommodations		
State/Province	**Park Name**	NPS Guided Tours	Hiking	Horseback Riding	Swimming	Boating	Fishing	Snowmobile Route	Cross-Country Ski Trail	Bathhouses	Groceries, Ice	Museum/Exhibit	Restaurant, Snacks	Handicap Access Rest Rooms	Cabins	Hotel, Motel, Lodge	Campgrounds
	Independence Natl. Historical Park, 313 Walnut St., Philadelphia, PA 19106	●										●	●	●			
	Upper Delaware Scenic and Recreational River (Pa., N.Y.), P.O. Box C, Narrowsburg, NY 12764	●		●	●	●	●				●	●	●	●	●	●	●
	Valley Forge Natl. Historical Park, P.O. Box 953, Valley Forge, PA 19481		●	●		●	●		●			●	●	●			
Prince Edward Island	Prince Edward Island Natl. Park, P.O. Box 487, Charlottetown, PE C1A 7L1	●	●		●	●	●		●	●		●	●	●			●
Québec	La Mauricie Natl. Park, 794 5th St., P.O. Box 758, Shawinigan, PQ G9N 6V9	●	●		●	●	●		●	●	●	●	●	●		●	●
South Carolina	Congaree Swamp Natl. Monument, 200 Caroline Sims Rd., Hopkins, SC 29061	●	●			●	●							●			
	Fort Sumter Natl. Monument, 1214 Middle St., Sullivans Island, SC 29482	●					●					●		●			
	Ninety Six Natl. Historic Site, P.O. Box 496, Ninety Six, SC 29666	●	●	●			●					●		●			
South Dakota	Badlands Natl. Park, P.O. Box 6, Interior, SD 57750	●	●									●	●		●		●
	Jewel Cave Natl. Monument, R.R. 1, Box 60AA, Custer, SD 57730	●										●		●			
	Mount Rushmore Natl. Memorial, P.O. Box 268, Keystone, SD 57751											●	●	●			
	Wind Cave Natl. Park, Hot Springs, SD 57747	●	●									●	●	●			●
Tennessee	Andrew Johnson Natl. Historic Site, P.O. Box 1088, Greeneville, TN 37744											●		●			
	Great Smoky Mountains Natl. Park (Tenn., N.C.), Gatlinburg, TN 37738	●	●	●			●				●	●		●	●	●	●
Texas	Amistad Recreation Area, P.O. Box 420367, Del Rio, TX 78842-0367		●		●	●	●				●		●	●			●
	Big Bend Natl. Park, Big Bend Natl. Park, TX 79834	●	●	●		●	●				●	●	●	●		●	●
	Fort Davis Natl. Historic Site, P.O. Box 1456, Fort Davis, TX 79734		●									●		●			
	Guadalupe Mountains Natl. Park, H.C. 60, Box 400, Salt Flat, TX 79847	●	●									●		●			●
	Lake Meredith Recreation Area, P.O. Box 1438, Fritch, TX 79036				●	●	●				●	●					●
	Lyndon B. Johnson Natl. Historical Park, P.O. Box 329, Johnson City, TX 78636	●										●		●			
	Padre Island Natl. Seashore, 9405 S. Padre Island Dr., Corpus Christi, TX 78418-5597		●	●	●	●	●			●		●		●			●
	San Antonio Missions Natl. Historical Park, 2202 Roosevelt Ave., San Antonio, TX 78210	●	●									●					
Utah	Arches Natl. Park, P.O. Box 907, Moab, UT 84532	●	●									●		●			●
	Bryce Canyon Natl. Park, Bryce Canyon, UT 84717	●	●	●					●		●	●	●	●	●	●	●
	Canyonlands Natl. Park, 125 West 200 South, Moab, UT 84532	●	●	●		●								●			●
	Capitol Reef Natl. Park, Torrey, UT 84775	●	●	●			●					●		●			●
	Cedar Breaks Natl. Monument, P.O. Box 749, Cedar City, UT 84720	●	●					●	●			●		●			●
	Glen Canyon Natl. Recreation Area (Utah, Ariz.), P.O. Box 1507, Page, AZ 86040	●	●		●	●	●				●	●	●	●		●	●
	Natural Bridges Natl. Monument, Box 1, Lake Powell, UT 84533		●									●					●
	Timpanogos Cave Natl. Monument, R.R. 3, Box 200, American Fork, UT 84003	●										●	●				
	Zion Natl. Park, Springdale, UT 84767-1099	●	●	●				●	●			●	●	●	●	●	●
Virginia	Appomattox Court House Natl. Historical Park, P.O. Box 218, Appomattox, VA 24522	●	●									●		●			
	Booker T. Washington Natl. Monument, Rte. 3, Box 310, Hardy, VA 24101	●	●									●					
	Colonial Natl. Historical Park, P.O. Box 210, Yorktown, VA 23690	●	●									●		●			
	George Washington Birthplace Natl. Mon., R.R. 1, Box 717, Washington's Birthplace, VA 22443	●	●				●					●		●			
	George Washington Memorial Pkwy. (Va., Md.), Turkey Run Park, McLean, VA 22101	●	●	●		●	●					●	●	●			
	Shenandoah Natl. Park, Rte. 4, Box 348, Luray, VA 22835	●	●	●			●		●		●	●	●	●	●	●	●
	Wolf Trap Farm for the Performing Arts, 1551 Trap Rd., Vienna, VA 22182	●											●	●			
Washington	Coulee Dam Natl. Recreation Area, P.O. Box 37, Coulee Dam, WA 99116	●			●	●	●			●		●		●			●
	Fort Vancouver Natl. Historic Site, 612 E. Reserve St., Vancouver, WA 98661-3897	●										●		●			
	Lake Chelan Natl. Recreation Area, 2105 WA 20, Sedro Woolley, WA 98284	●	●	●		●	●		●		●		●		●	●	●
	Mount Rainier Natl. Park, Tahoma Woods, Star Route, Ashford, WA 98304	●	●				●	●	●		●	●	●	●		●	●
	North Cascades Natl. Park, 2105 WA 20, Sedro Woolley, WA 98284		●	●			●										
	Olympic Natl. Park, 600 E. Park Ave., Port Angeles, WA 98362	●	●	●	●	●	●		●	●	●	●	●	●	●	●	●
West Virginia	Harpers Ferry Natl. Historical Park (W.Va, Md., Va), P.O. Box 65, Harpers Ferry, WV 25425	●	●				●					●		●			
Wisconsin	Apostle Islands Natl. Lakeshore, Rte. 1, Box 4, Bayfield, WI 54814	●	●		●	●	●					●		●			●
	St. Croix and Lower St. Croix Natl. Scenic Riverways (Wis., Minn.), P.O. Box 708, St. Croix Falls, WI 54024		●		●	●	●		●		●	●	●	●	●	●	●
Wyoming	Devils Tower Natl. Monument, Devils Tower, WY 82714	●	●				●					●		●			●
	Fort Laramie Natl. Historic Site, Fort Laramie, WY 82212	●					●					●		●			
	Grand Teton Natl. Park, P.O. Drawer 170, Moose, WY 83012	●	●	●	●	●	●	●	●		●	●	●	●	●	●	●
	Yellowstone Natl. Park (Wyo., Idaho, Mont.), P.O. Box 168, Yellowstone Natl. Park, WY 82190	●	●	●		●	●	●	●		●	●	●	●	●	●	●

Tourism Information & Auto/RV Laws

Cars Towing Trailers
- Riding in Trailer Permitted
- Flares Required
- Chains Required
- Max Length of Car and Trailer Without Permit
- Breakaway Brakes if Weight Over
- Brakes if Weight Over
- Overnight Off-road Parking Permitted

Automobiles
- Auto Liability Insurance Mandatory
- Seat Belts Required
- Child Restraints Required
- Studded Tires Permitted (Dates)

State	Tourism Information	Studded Tires Permitted (Dates)	Child Restraints Required	Seat Belts Required	Auto Liability Insurance Mandatory	Overnight Off-road Parking Permitted	Brakes if Weight Over	Breakaway Brakes if Weight Over	Max Length of Car and Trailer Without Permit	Chains Required	Flares Required	Riding in Trailer Permitted
Alabama	Alabama Bureau of Tourism & Travel 800/ALABAMA; 205/242-4169	Prohibited	•	•		NO	3,000	3,000	85			
Alaska	Alaska Division of Tourism 907/465-2010	Oct. 1–Apr. 15 (d)	•	•	•	Designated areas only	(b)	(b)	48	•		
Arizona	Arizona Office of Tourism 602/542-TOUR	Oct. 1–May 1 (d)	•	•	•	Only if posted	3,000	3,000	40	•		
Arkansas	Arkansas Dept. of Parks & Tourism 800/NATURAL	Nov. 15–Apr. 15	•	•	•	Rest areas only	3,000	3,000	65	•		
California	California Office of Tourism 800/TO-CALIF, ext. A1003	Nov. 1–Apr. 1	•	•	•	NO	3,000	3,000	65	•	• d	• d
Colorado	Colorado Tourism Board 800/433-2656; 303/592-5510	YES	•	•	•	Unless posted	3,000	3,000	70	•		
Connecticut	Connecticut Dept. of Economic Development 800/CT-BOUND; 203/258-4356	Nov.1–Apr. 15	•	•	•	Rest areas only	3,000	3,000	60	•		
Delaware	Delaware Tourism Office 800/441-8846	Oct.15–Apr. 15	•	•	•	NO	4,000	4,000	65		• d	
District of Columbia	Washington Conv. & Visitors Assn. 202/789-7000	Oct. 15–Apr. 15	•	•		NO	3,000(a)	3,000	55	•		
Florida	Florida Division of Tourism 904/487-1462	Prohibited	•	•	•	YES, unless posted	3,000	3,000	50	•	•	
Georgia	Georgia Dept. of Tourism 800/VISIT GA; 404/656-3590	Prohibited	•	•	•	Designated areas only			60			
Hawaii	Hawaii Visitors Bureau 808/923-1811	Prohibited	•	•	•	NO	3,000	3,000	45	•		
Idaho	Idaho Division of Tourism Development 800/635-7820; 208/334-2470	Oct. 1–Apr. 15	•	•	•	Designated areas only	1,500	1,500	48		• d	
Illinois	Illinois Bureau of Tourism 800/223-0121	Prohibited	•	•	©	YES	3,000	5,000	60	• d		
Indiana	Indiana Tourism Division 317/232-8860; 800/289-ONIN	Oct. 1–May 1	•	•	•	NO	(b)	3,000	60	•	•	•
Iowa	Iowa Division of Tourism 800/345-IOWA; 515/242-4705	Nov. 1–Apr. 1	•	•		Not at rest areas	3,000	3,000	60	•	•	•
Kansas	Kansas Travel & Tourism 800/2 KANSAS; 913/296-2009	Nov. 1–Apr. 15 (d)	•	•	•	YES	(b,d)		65	• d		
Kentucky	Kentucky Dept. of Travel Development 800/225-TRIP; 502/564-4930	YES	•		•	NO	(b)		55			•
Louisiana	Louisiana Office of Tourism 800/33GUMBO; 504/342-8119	Prohibited	•	•	•	NO	1,500	3,000	65	•	•	
Maine	Maine Publicity Bureau 800/533-9595; 207/582-9300	Oct. 1–May 1	•		•	YES	3,000		65			
Maryland	Maryland Office of Tourism Development 800/543-1036; 410/333-6611	Prohibited	•	•	•	Designated areas only	3,000	3,000	35	•		
Massachusetts	Massachusetts Office of Travel & Tourism 800/447-MASS; 617/727-3201	Nov. 1–Apr. 1	•		•	Overnight rest areas only	10,000	10,000	66	•	• d	
Michigan	Michigan Dept. of Commerce 800/5432-YES; 517/373-0670	Prohibited	•	•	•	Unless posted	5,500				•	• d
Minnesota	Minnesota Dept. of Tourism 800/657-3700; 612/296-5029	Prohibited (d)	•	•	•	Not where posted or on Freeways	3,000	6,000	65	•	•	•
Mississippi	Mississippi Div. of Tourism 800/647-2290; 601/359-3297	Prohibited	•	•		NO	2,000		50 for trailer	•		•
Missouri	Missouri Div. of Tourism 314/751-4133; 800/877-1234	Nov. 1–Mar. 31	•	•	•	YES			55			•

(a) Or, when trailer exceeds 40% of weight of towing vehicle   (b) Equipment required   (c) Must be able to stop within legal distance
(d) Possible exceptions. Refer to state laws   (e) Studded tires permitted if studs don't project more than 1⁄16" when compressed

Cars Towing Trailers

- Riding in Trailer Permitted
- Flares Required
- Chains Required
- Max Length of Car and Trailer Without Permit
- Breakaway Brakes if Weight Over
- Brakes if Weight Over
- Overnight Off-road Parking Permitted

Automobiles

- Auto Liability Insurance Mandatory
- Seat Belts Required
- Child Restraints Required
- Studded Tires Permitted (Dates)

State	Tourism Information	Studded Tires Permitted (Dates)	Child Restraints Req.	Seat Belts Req.	Auto Liability Ins.	Overnight Off-road Parking	Brakes if Weight Over	Breakaway Brakes if Weight Over	Max Length	Chains Req.	Flares Req.	Riding in Trailer
Montana	Travel Montana 800/541-1447; 406/444-2654	Oct. 1–May 31	•	•	•	NO	3,000	3,000	75	•(d)	•(d)	d
Nebraska	Nebraska Travel & Tourism 800/228-4307; 402/471-3796	Nov. 1–Apr. 1	•	•	•	Not in rest areas	3,000	3,000	65		•(d)	•
Nevada	Nevada Commission on Tourism 800/NEVADA8 (U.S. & W. Canada)	Oct. 1–Apr. 30	•	•	•	YES	1,500(d)	3,000	70		•(d)	
New Hampshire	New Hampshire Office of Vacation Travel & Tourism 603/271-2666	YES, with no restrictions	•			YES	3,000(d)	3,000	48	•	•(d)	
New Jersey	New Jersey Division of Travel & Tourism 800/JERSEY-7; 609/292-2470	Nov. 15–Apr. 1	•	•	•	Rest areas only	3,000		50	•	•	•
New Mexico	New Mexico Dept. of Tourism 800/545-2040; 505/827-7777	YES, with no restrictions	•	•	•	NO			40	•	•	
New York	New York Div. of Tourism 800/CALL-NYS; 518/474-4116	Oct. 1–May 14	•	•	•	YES	1,000		65	•		
North Carolina	North Carolina Travel & Tourism Div. 800/VISIT-NC; 919/733-4171	YES, with no restrictions	•	•	•	NO	4,000		60			•
North Dakota	North Dakota Parks & Tourism 800/435-5663; 701/224-2525	Oct. 15–Apr. 15	•		•	NO (d)			75	•		d
Ohio	Ohio Div. of Travel & Tourism 800/BUCKEYE; 614/466-8844	Nov. 1–Apr. 15	•	•	•	Where posted	2,000	2,000	65	•		
Oklahoma	Oklahoma Tourism & Recreation Dept. 800/652-6552; 405/521-2406	Nov. 1–Apr. 1	•	•	•	Only at rest areas	3,000	3,000	70	•	•	•
Oregon	Oregon Tourism Division 800/547-7842; 503/378-3451	Nov. 1–Apr. 30	•	•	•	YES, if no hazard			50	•		d
Pennsylvania	Pennsylvania Bureau of Travel Marketing 800/VISIT-PA	Nov. 1–Apr. 15	•	•		Posted areas only	(b)	3,000	60	•	•	
Rhode Island	Rhode Island Tourist Div. 800/556-2484; 401/277-2601	Nov. 15–Mar. 31	•	•	•	Designated rest areas only	(b)	3,000		•		
South Carolina	South Carolina Dept. of Parks, Recreation & Tourism 803/734-0235	YES (e)	•	•	•	Designated areas only	3,000	3,000	48 for trailer	•		
South Dakota	South Dakota Tourism 605/773-3301	Oct. 1–Apr. 30	•		•	Designated areas only	3,000	3,000	80	•	•(d)	d
Tennessee	Tennessee Tourist Dev. 615/741-2158	Oct. 1–Apr. 15	•	•		NO	3,000	3,000		•		•
Texas	Texas Tourist Division 800/8888 TEX (out of state) 800/452-9292 (in state)	Prohibited	•	•	•	YES, unless posted	4,500	3,000	65	•(d)	•(d)	d
Utah	Utah Travel Council 801/538-1030	Oct. 15–Mar. 31	•	•	•	YES	2,000	3,000	65	•	•	
Vermont	Vermont Travel Division 800/338-0189 (out of state) 802/828-3236 (in state)	YES	•		•	NO	3,000(a)	3,000	65	•		•(d)
Virginia	Virginia Tourism Development 800/932-5827; 804/786-4484	Oct. 15–Apr. 15	•	•		NO	3,000(a)	(b)	60	•		d
Washington	Washington Tourism Dev. 800/544-1800; 206/586-2102	Nov. 1–Apr. 1	•	•	•	NO	(b)	3,000	40	•	•	
West Virginia	West Virginia Div. of Tourism & Parks 800/CALL-WVA	Nov. 1–Apr. 15	•		•	Not on Interstates	3,000	3,000	55	•	•	•
Wisconsin	Wisconsin Division of Tourism 800/432-TRIP; 608/266-2161	Prohibited	•	•		NO	3,000	(b)	65	•		d
Wyoming	Wyoming Div. of Tourism 800/CALL-WYO (out of state) 307/777-7777 (in state)	YES, with no restrictions	•	•	•	Designated areas only	(b,c)		85		•	d

(a) Or, when trailer exceeds 40% of weight of towing vehicle (b) Equipment required (c) Must be able to stop within legal distance
(d) Possible exceptions. Refer to state laws (e) Studded tires permitted if studs don't project more than 1/16" when compressed

©1993 Rand McNally & Company

Canada Travel Information

Canadian Citizens Visiting the United States

Canadian nationals, and aliens having a common nationality with nationals of Canada, are not required to present passports or visas to visit for a period of six months or less, except when arriving from a visit outside the Western Hemisphere. However, such persons should carry evidence of their citizenship. Visitors entering for a period of more than six months and less than one year are required to furnish valid passports.

This article does not attempt to cover all regulations or requirements. Obtain further information from the nearest United States or Canadian customs office in the U.S., Canada, or a port of entry.

United States Citizens Visiting Canada

Passports are not required of native-born United States citizens to enter Canada. They should carry identifying papers such as a passport, certified birth certificate plus a photo ID, or a voter's registration card plus a photo ID with the same address to show proof of their U.S. citizenship at the port of entry. Naturalized citizens must carry their naturalization certificate. Aliens who reside in the United States must have their Alien Registration Receipt Card.

Automobiles will be admitted for touring in Canada without payment of any duty or fee for any period up to twelve months. Vehicle Registration Cards should be carried. Any necessary permits are issued at any port of entry.

Returning motorists must report for inspection. Each U.S. resident, after 48 hours, may bring back articles for personal use valued at up to $400.00 free of duty, provided same has not been claimed in the preceding 30 days.

Hunting in Canada

Hunting is controlled by federal, provincial, and territorial laws. Hunting licenses obtained from each province or territory in which they plan to hunt are required of all non-residents. An additional permit, the federal migratory game bird hunting permit, is also required for those planning to hunt migratory game birds. Obtain the permit from:

Environment Canada
Canadian Wildlife Service
Ottawa, Ontario, Canada K1A 0H3
819/997-2957

Many of Canada's provincial parks and reserves forbid the entry of any type of weapon. Each province can provide you with regulations. Export permits are required to take out all unprocessed wildlife from the Northwest Territories.

The importation of firearms into Canada is strictly controlled. You may bring a hunting rifle or shotgun and up to 200 rounds of ammunition into Canada for sporting or competition use, but you must be at least 16 years of age (19 in British Columbia). No hand guns or automatic weapons are allowed. For specific rules and further information:

Revenue Canada, Customs and Excise
Commercial Verification & Enforcement
Connaught Building, Mackenzie Ave.
Ottawa, Ontario, Canada K1A 0L5
613/995-3331

Fishing in Canada

Fishing is likewise controlled by federal, provincial, and territorial laws. Fishing licenses obtained from each province or territory in which they plan to fish are required of all non-residents. British Columbia also requires tidal-waters sports fishing licenses.

Fishing in national wildlife areas is controlled by Environment Canada (see address above). Special fishing permits are required to fish in all national parks. These permits are available at any national park site for a small fee. For information on rules in national parks write to:

National Parks of Canada
Environment Canada
Ottawa, Ontario, Canada K1A 0H3

Hunting & Fishing in the United States

Hunting and fishing are controlled by the respective states. Information regarding licenses, seasons, fees, and specific rules may be obtained by contacting the individual states in which you plan to hunt or fish. (See Fishing/Hunting chart for addresses of state agencies.)

Firearms and Ammunition

Firearms and ammunition are subject to restrictions and import permits approved by the Bureau of Alcohol, Tobacco and Firearms (ATF). In order to import these items, applications to import may be made only by or through a licensed importer, dealer, or manufacturer. Items prohibited by the National Firearms Act will not be admitted unless specifically authorized.

If a person is returning with firearms or ammunition that was previously taken out of the U.S. by that person, no import permit is required upon presentation of proof of such action. To facilitate reentry, have the items registered, before departing from the United States, at any Customs office or ATF field office. No more than three nonautomatic firearms and 1,000 cartridges will be registered for any one person.

For further information, contact:
Alcohol, Tobacco & Firearms
Department of the Treasury
Washington, D.C. 20226

CANADIAN TOURISM OFFICES

If you wish to obtain information about particular areas, places of interest, activities, or events, contact the provincial or territorial offices directly.

Alberta Tourism, Parks & Recreation
10155-102 Street
Edmonton, AB T5J 4L6
800/661-8888 (N. Am. exc. AB, AK, & HI)
800/222-6501 (in province)
403/427-4321

Tourism **British Columbia**
1117 Wharf St.
Victoria, BC V8W 2Z2
800/663-6000 (N. Am.)
604/387-1642

Manitoba Department of Industry, Trade & Tourism
7-155 Carlton St.
Winnipeg, MB R3C 2H8
800/665-0040 ext. 275 (N. Am.)
204/945-3777

New Brunswick Dept. of Economic Growth & Tourism
P.O. Box 12345
Fredericton, NB E3B 5C3
800/561-0123 (N. Am.)
800/442-4442 (in province)

Newfoundland and **Labrador** Tourism
P.O. Box 8700
St. John's, NF A1B 4J6
800/563-6353 (N. Am.)
709/729-2830

Northwest Territories Dept. of Economic Development & Tourism
Box 1320
Yellow Knife, NT X1A 2L9
800/661-0788 (N. Am.)
403/873-7200

Nova Scotia Dept. of Tourism & Culture
Box 456
Halifax, NS B3J 2R5
800/565-0000 (Canada)
800/341-6096 (U.S. exc AK, HI)
902/424-4207

Ontario Travel
Queen's Park
Toronto, ON M7A 2E5
800/ONTARIO (N. Am. exc AK)
416/314-0944

Prince Edward Island Dept. of Tourism, Parks, & Recreation
P.O. Box 940
Charlottetown, PE C1A 7N8
800/565-0267 (N. Am. exc AK, HI)
902/368-4444

Tourisme **Québec**
P.O. Box 20,000
Québec, PQ G1W 2E2
800/363-7777 (N. Am. exc AK)
514/873-2015

Tourism **Saskatchewan**
1919 Saskatchewan Dr.
Regina, SK S4P 3V7
800/667-7191 (N. Am.)
306/787-2300

Tourism **Yukon**
Box 2703
Whitehorse, YT Y1A 2C6
403/667-5340

General information about Canada is available from Canadian Consulates in major U.S. cities.

Road Condition "Hot Lines"

Call the following numbers for road conditions and road construction.

Alabama
conditions: (205) 242-4378 24 hours

Alaska
conditions: (907) 243-7675, #2 (recording)
construction: (907) 243-7675, #8 (recording)

Arizona
conditions & construction: (602) 252-1010, ext. 7623 (recording)
construction: (602) 255-7386 weekdays

Arkansas
conditions: (501) 569-2374 (recording)
construction: (501) 569-2227 weekdays

California
conditions & construction: (916) 653-7623 (recording)

Colorado
conditions: (303) 639-1234 (recording)
construction: (303) 757-9228 weekdays

Connecticut
conditions & construction: (800) 443-6817 (CT only)
conditions: (203) 566-4880 weekdays

Delaware
conditions: (302) 739-4313 weekdays
construction: (302) 674-1441 (recording)

District of Columbia
conditions: (202) 936-1111 (recording)
construction: (202) 939-8099 weekdays

Florida
no central source

Georgia
conditions: (800) 722-6617 (GA only);
(404) 656-5267 weekdays

Hawaii
construction: (808) 536-6566 (recording)

Idaho
conditions: (208) 336-6600 (recording)
construction: (208) 334-8888 (recording)

Illinois
conditions: (312) 368-4636 (recording);
(217) 782-5730 (recording in winter)

Indiana
conditions—north: (317) 232-8300 (recording)
conditions—south: (317) 232-8298 (recording)
construction: (317) 232-5115 weekdays

Iowa
conditions: (515) 288-1047 (recording)
construction: (515) 239-1471 weekdays

Kansas
conditions & construction: (913) 296-3102
24 hours

Kentucky
conditions: (502) 564-4556 weekdays
construction: (502) 564-3730 weekdays

Louisiana
conditions & construction: (504) 379-1541
weekdays

Maine
conditions Nov.–Apr. (24 hrs); May-Oct. (weekdays): (207) 289-3427
construction: (207) 289-3171 weekdays

Maryland
conditions: (800) 543-2515 (MD only);
(401) 333-1215
construction: (800) 222-5943 (MD only);
(410) 333-1122 by county/weekdays

Massachusetts
conditions & construction: (617) 973-7500
weekdays

Michigan
conditions & construction: (517) 973-7500
24 hours

Minnesota
conditions: (800) 542-0220 (recording);
(612) 296-3076 (recording)

Mississippi
conditions: (601) 987-1212 24 hours

Missouri
No central source

Montana
conditions: (800) 332-6171 (recording);
(406) 444-6339 (recording)

Nebraska
conditions & construction: (402) 479-4512
weekdays
conditions (winter): (402) 471-4533 (recording)

Nevada
conditions & construction recordings for:
South—Las Vegas: (702) 486-3116
Northwest—Reno: (702) 793-1313
Northeast—Elko: (702) 738-8888

New Hampshire
conditions & construction: (603) 485-3851
24 hours

New Jersey
conditions & construction for the Turnpike:
(908) 247-0900 24 hours
conditions & construction for the Garden State Parkway: (908) 727-5929 (recording)

New Mexico
conditions: (505) 827-5213 weekdays
then recording)
construction: (800) 432-4269 (NM only)
weekdays; (505) 827-5118 weekdays

New York
conditions: (800) 843-7623 (NY only) (recording)
conditions in Canada & Northeastern states:
(800) 247-7204 (recording)

North Carolina
conditions: (919) 733-3861 24 hours

North Dakota
conditions: (800) 472-2686 (ND only) (recording);
(701) 224-2898 (recording)
construction: (701) 224-4418 weekdays

Ohio
conditions & construction: (614) 466-7170
weekdays

Oklahoma
construction: (800) 522-7623 (OK only) (recording)

Oregon
conditions & construction: (800) 976-7277
(OR only); (503) 889-3999 (recording)

Pennsylvania
conditions & construction for the Turnpike:
(800) 331-3414 (PA only)
conditions: (717) 939-9551 ext. 5550 weekdays
conditions Nov.–May: (717) 939-9871 (recording)
conditions for the Interstate Oct.–Apr.:
(814) 355-7545

Rhode Island
conditions: (401) 738-1211 (recording)
construction: (401) 277-2468 weekdays

South Carolina
conditions & construction: (803) 737-1030
24 hours

South Dakota
conditions & construction: (605) 773-3536
24 hours

Tennessee
conditions: (615) 251-5227 weekdays

Texas
conditions: (512) 463-8588 weekdays;
(then recording)

Utah
conditions & construction: (801) 964-6000
(recording)

Vermont
conditions & construction: (802) 828-2468
weekdays

Virginia
conditions & construction: (800) 367-ROAD
(VA only)
conditions (winter): (804) 786-3181 weekdays

Washington
conditions: Mountain Pass Report (Nov.–Apr.)
1-976-ROAD; (206) 434-ROAD

West Virginia
conditions: (304) 348-3758 or -2889 (recording);
(304) 348-3028 24 hours

Wisconsin
conditions (winter) & construction (summer):
(800) 762-3947 (recording)

Wyoming
conditions: (307) 635-9966 (recording)
construction: (307) 777-4437 weekdays

Keys:
weekdays = normal business hours
24 hours = a person answers 24 hours/7 days per week
recording = available at all times unless noted

Intercity Toll Road Information

State	Road Name	Location	Miles	Auto Toll	Auto/2 Axle Trailer
Delaware	Kennedy Memorial Highway	Md. State Line to Wilmington	11.2	$1.00	$3.00
Florida	Bee Line Main	Orlando Airport Plaza to FL 520	20	1.00	2.00
	Bee Line East	FL 520 to Cape Canaveral	22	.20	.40
	Bee Line West	I-4 to Co 436 (Semoran Blvd.)	11	.50	1.00
	Everglades Parkway (Alligator Alley)	Naples to Andytown	78	1.50	4.00
	Florida's Turnpike	I-75 to Miami	260	12.50	26.00
	Florida's Turnpike (Homestead Extension)	Miramar to Florida City	48	2.00	6.00
	Sawgrass Expressway	I-75 to I-95	23	1.50	3.10
Illinois	Chicago Skyway	I-94, Chicago, to Ind. State Line	7.3	1.75	2.50
	East-West Tollway	I-88, Chicago, to Rock Falls	97	3.00	6.00
	North-South Tollway	I-290, Addison, to I-55, Bolingbrook	17.5	1.00	2.00
	Northwest Tollway	Des Plaines to South Beloit	76	2.00	4.00
	Tri-State Tollway	Ind. State Line to Wis. State Line	83	2.40	4.80
Indiana	Indiana Toll Road	Ohio State Line to Ill. State Line	157	4.65	5.35
Kansas	Kansas Turnpike	Kansas City to Okla. State Line	236	7.00	9.50
Kentucky	Audubon Parkway	Pennyrile Parkway to Owensboro	23	.50	1.10
	Cumberland Parkway	Bowling Green to Somerset	88.2	2.00	4.00
	Daniel Boone Parkway	London to Hazard	59.4	1.40	2.80
	Green River Parkway	Owensboro to Bowling Green	69.7	1.50	3.00
	Pennyrile Parkway	Hopkinsville to Henderson	60	1.00	2.30
	Purchase Parkway	Fulton to US 62 near Gilbertsville	49	.90	2.00
Maine	Maine Turnpike	York to Augusta	100	3.10	4.65
Massachusetts	Massachusetts Turnpike	Boston to N.Y. State Line	134.6	5.60	6.70
Mexico	Ensenada Toll Rd.	Tijuana to Ensenada	57	2.50**	5.00**
	Mexico City-Iguala Toll Rd.	Mexico City to Iguala	103	12.00**	24.00**
	Mexico Highway 57	Mexico City to Palmillas	85	10.50**	21.00**
	Mexico City-Puebla Toll Rd.	Mexico City to Puebla	65	5.50**	10.50**
	Mexico City-Toluca Toll Rd.	Mexico City to Toluca	41	6.00**	12.00**
New Hampshire	F.E. Everett Turnpike	Nashua to Concord	39	1.00	1.50
	New Hampshire Turnpike	Portsmouth to Seabrook	16	.75	1.25
	Spaulding Turnpike	Portsmouth to Rochester, N.H.	31	.50	1.00
New Jersey	Atlantic City Expressway	Turnersville to Atlantic City	44	1.25	3.75
	Garden State Parkway	Montvale to Cape May	173	3.85	6.00
	New Jersey Turnpike	Delaware Memorial Bridge to George Washington Bridge	142	4.60	15.60
New York	New York Thruway – Eastbound	Pa. State Line to N.Y.C.	496	18.10	33.30
	Westbound	N.Y.C. to Pa. State Line	496	15.60	28.80
	Berkshire Section	Selkirk to Mass. Turnpike	24	1.20	2.15
	New England Section	N.Y.C. to Conn. State Line	15	1.00	2.00
	Niagara Section	Buffalo to Niagara Falls	21	1.50	3.00
Ohio	James W. Shocknessy Ohio Turnpike	Pa. State Line to Ind. State Line	241	4.90	7.50
Oklahoma	Cherokee Turnpike	US 412 to US 59	33	1.50	3.00
	Cimarron Turnpike	I-35 to Tulsa	59.2	1.75	3.50
	H.E. Bailey Turnpike	Oklahoma City to Texas State Line	86.4	2.75	5.25
	Indian Nation Turnpike	Henryetta to Hugo	105.2	3.50	6.25
	Kilpatrick Turnpike	Oklahoma City, OK 74 to I-35/44	8	.75	1.50
	Muskogee Turnpike	Tulsa to Webber Falls	53.1	1.75	3.25
	Turner Turnpike	Oklahoma City to Tulsa	86	2.50	5.25
	Will Rogers Turnpike	Tulsa to Mo. State Line	88.5	2.50	5.25
Pennsylvania	Beaver Valley Expressway	New Castle to Beaver Falls	17	.50	2.00
	Pennsylvania Turnpike	N.J. State Line to Ohio State Line	358	14.70	21.70
	Pa. Turnpike (N.E. Sect.)	Norristown to Scranton	110	4.15	6.25
Texas	Dallas North Tollway	I-35E, Dallas, to FM 544, Dallas	17.2	1.00	1.60
	Hardy Toll Road	I-45, Houston, to I-610, Houston	21.7	1.00*	5.00
	Sam Houston Tollway	US 59, Houston, to I-45	28	2.25	5.25
Virginia	Chesapeake Bay Bridge & Tunnel	US 13, Norfolk/VA Beach to Eastern Shore	17	9.00	15.00
	Dulles Toll Road	VA 123 to VA 28	13	.85	1.70
	Powhite Parkway	I-195 to Old Hundred Rd.	12.5	1.10	1.70
	Richmond – Petersburg Tpk.	Richmond to Petersburg	35	1.50	2.10
	Virginia Bch. – Norfolk Expy.	US 60, Virginia Bch., to I-64, Norfolk	12.1	.25	.40
West Virginia	West Virginia Turnpike	Charleston to Princeton	88	3.75	6.00

*75¢ with correct change   **At 1992 rate of exchange   †Less than $2.00

©1993 Rand McNally & Company

Car Rental Companies Toll-free Numbers

The following selected list of car rental companies provides toll-free "800" numbers for making reservations in Canada and the United States. Although these numbers were in effect at press time, the Guide cannot be responsible should any of these numbers change. Many companies do not have toll-free reservation numbers; consult your local telephone directory for a regional listing.

Toll-free Numbers

AAPEX Courtesy Car Rental
Lighthouse Point, FL 33064
3400 N. Federal Hwy.
(800) 327-9106 Cont'l USA, except FL, and Canada

Agency Rent-A-Car
Solon, OH 44139
Corporate Office
30000 Aurora Rd.
(800) 321-1972 Cont'l USA and Canada

Alamo Rent-A-Car
Ft. Lauderdale, FL 33335
Corporate Office
P.O. Box 22776
(800) 327-9633 USA and Canada

Allstate Car Rental
Las Vegas, NV 89119
McCarren Int'l Airport
5175 Rent-A-Car Rd.
(800) 634-6186 USA and Canada

Altra Auto Rental
Solon, OH 44139
Corporate Office
30000 Aurora Rd.
(800) 232-9555 Cont'l USA except OH
(800) 621-9184 OH

American International Rent-A-Car
East Boston, MA 02128
Mass Tech Center
One Harborside Dr.
(800) 527-0202 USA and Canada

Avis-Reservations Center
Garden City, NY 11530-9795
900 Old Country Rd.
(800) 331-1212 Domestic Reservations
(800) 331-1084 Int'l Reservations, including Canada

Aztec Car Rental
San Diego, CA 92101
2401 Pacific Hwy.
(800) 231-0400 Cont'l USA and Canada

Brooks Car Rental
Las Vegas, NV 89109
3765 Las Vegas Blvd. South
(800) 634-6721 Cont'l USA

Budget Rent-A-Car
Chicago,IL 60601
200 N. Michigan Ave.
(800) 527-0700 Cont'l USA and Canada
(800) 472-3325 Int'l Reservations

Dollar Rent-A-Car
Los Angeles, CA 90045
6141 W. Century Blvd.
(800) 800-4000 USA and Canada

Enterprise Rent-A-Car
St. Louis, MO 63124
8850 Ladue Rd.
(800) 325-8007 Cont'l USA and Canada

Fairway Rent-A-Car
Las Vegas, NV 89109-1922
3469 Industrial Rd.
(800) 634-3476 USA and Canada

Freedom Rent-A-Car
P.O. Box 2345
705-B Yucca
Boulder City, NV 89005
(800) 331-0777 Cont'l USA and Canada

General Rent-A-Car
Hollywood, FL 33020
2741 N. 29th Ave.
(800) 327-7607 Cont'l USA

Hertz Corporation
Oklahoma City, OK 73120
Worldwide Reservation Center
10401 N. Pennsylvania
(800) 654-3131 Cont'l USA
(800) 654-3001 International

Interamerican Car Rental
Miami, FL 33126
1790 NW. Le Jeune Rd.
(800) 327-1278 Cont'l USA

National Car Rental
Minneapolis, MN 55435
7700 France Ave. South
(800) CAR-RENT Cont'l USA and Canada
(800) CAR-EURO International

Payless Car Rental Int'l Inc.
St. Petersburg, FL 33784-0669
P.O. Box 60669
(800) PAYLESS USA and Canada

Sears Rent-A-Car
Chicago, IL 60601
200 N. Michigan Ave.
(800) 527-0770 USA and Int'l Reservations

Showcase Rent-A-Car
Los Angeles, CA 90045
9220 S. Sepulveda Blvd.
(800) 421-6808 USA and Canada

Thrifty Rent-A-Car
Tulsa, OK 74153-0250
P.O. Box 35250
(800) 367-2277 USA and Canada

USA Rent-A-Car System
Tampa, FL 33607
4350 W. Cypress St., Suite 750
(800) 872-2277 USA

U-SAVE Auto Rental, Inc.
Hanover, MD 21076
7525 Connelley Dr., Suite A
(800) 272-U-SAV Cont'l USA

Value Rent-A-Car
Boca-Raton, FL 33431
P.O. Box 5040
(800) GO-VALUE Cont'l USA and Canada

Limousine Services

The following limousine service companies provide toll-free "800" numbers for making reservations throughout the United States.

Carey Limousine
Washington, DC 20016
4530 Wisconsin Ave., NW
(800) 336-4646 USA
(800) 336-4747 Canada

Dav-El Limousine
New York, NY 10011
North River, Pier 62
(800) 922-0343 USA and Canada

Hotel/Motel Toll-free Numbers

This selected list is a handy guide for hotel/motel toll-free reservation numbers. You can save time and money by using these toll-free numbers for continental USA and Canada. Although these "800" numbers were in effect at press time, the Atlas cannot be responsible should any of these numbers change. Many establishments do not have toll-free reservation numbers. Consult your local phone directory for a regional listing.

Adam's Mark Hotels
(800) 444-ADAM Cont'l USA and Canada

Best Value Inns/Sundowner/Superior
(800) 322-8029 Cont'l USA and Canada

Best Western International, Inc.
(800) 528-1234 USA and Canada

Budgetel Inns
(800) 4 BUDGET Cont'l USA

Canadian Pacific Hotels
(800) 828-7447 Cont'l USA
(800) 268-9411 Canada

Clarion Hotels
(800) CLARION Cont'l USA and AK
(800) 458-6262 Canada

Comfort Inns
(800) 228-5150 Cont'l USA and Canada

Country Hearth Inn
(800) 848-5767 Cont'l USA

Courtyard by Marriott
(800) 321-2211 Cont'l USA and Canada

Days Inn
(800) 325-2525 Cont'l USA and Canada

Delta Inns & Resorts
(800) 877-1133 USA
(800) 268-1133 Canada

Doubletree Hotels
(800) 528-0444 Cont'l USA and Canada

Downtowner/Passport Motor Inns
(800) 238-6161 Cont'l USA and Canada

Drury Inn
(800) 325-8300 Cont'l USA and Canada

Econo Lodges
(800) 55-ECONO Cont'l USA and Canada

Embassy Suites
(800) EMBASSY Cont'l USA and HI
(800) 458-5848 Canada

Exel Inns of America
(800) 356-8013 Cont'l USA and Canada

Fairfield Inns by Marriott
(800) 228-2800 Cont'l USA and Canada

Fairmont Hotels
(800) 527-4727 Cont'l USA and Canada

Forte Hotels
(800) 225-5843 Cont'l USA and Canada

Four Seasons Hotels & Resorts
(800) 332-3442 Cont'l USA
(800) 268-6282 Canada

Friendship Inns
(800) 453-4511 Cont'l USA and Canada

Guest Quarters Suite Hotel
(800) 424-2900 Cont'l USA and Canada

Hampton Inns
(800) HAMPTON Cont'l USA and Canada

Harley Hotels
(800) 321-2323 Cont'l USA and Canada

Helmsley Hotels
(800) 221-4982 Cont'l USA and Canada

Hilton Hotels
(800) HILTONS USA and Canada

Holiday Inns, Inc.
(800) HOLIDAY USA and Canada

Homewood Suites
(800) CALL-HOM(E) Cont'l USA and Canada

Hospitality International/Master Hosts/Red Carpet/Scottish Inns
(800) 251-1962 Cont'l USA

Howard Johnson Lodges
(800) 654-2000 Cont'l USA and Canada

Hyatt Hotels Corp.
(800) 233-1234 USA (except AK, NE) and Canada
(800) 228-9005 AK
(800) 228-3336 NE

Inter-Continental Hotels
(800) 327-0200 Cont'l USA, HI, and Canada

Journey's End Hotels
(800) 668-4200 USA and Canada

Knights Inn
(800) 843-5644 Cont'l USA and Canada

LK Motels
(800) 282-5711 Cont'l USA

La Quinta Motor Inns, Inc.
(800) 531-5900 Cont'l USA and Canada

Loews Hotels
(800) 23-LOEWS USA and Canada

Luxbury Hotels
(800) CLASS-4-U USA and Canada

Marriott Hotels & Resorts
(800) 228-9290 Cont'l USA, HI, and Canada

Master Hosts (See Hospitality International entry)

Meany Tower Hotels
(800) 648-6440 Cont'l USA

Meridien Hotels
(800) 543-4300 Cont'l USA and Canada

Omni Hotels
(800) THE OMNI USA and Canada

Preferred Hotels
(800) 323-7500 USA and Canada

Quality Inns
(800) 228-5151 Cont'l USA, HI, and Canada

Radisson Hotels International
(800) 333-3333 USA and Canada

Ramada Inns, Inc.
(800) 228-2828 USA and Canada

Red Carpet (See Hospitality International entry)

Red Lion Inns
(800) 547-8010 USA and Canada

Red Roof Inns
(800) THE ROOF USA and Canada

Regent International Hotels
(800) 545-4000 USA and Canada

Residence Inn by Marriott
(800) 331-3131 Cont'l USA and Canada

Relax Hotels and Resorts
(800) 667-3529 USA and Canada

Restcorp International
(800) 873-2392 Cont'l USA and Canada

The Ritz-Carlton
(800) 241-3333 USA and Canada

Rodeway Inns International
(800) 228-2000 Cont'l USA and Canada

Scottish Inns (See Hospitality International entry)

Sheraton Hotels & Motor Inns
(800) 325-3535 USA and Canada

Shoney's Inn
(800) 222-2222 USA
(800) 233-4667 Canada

Sonesta Hotels
(800) SONESTA USA and Canada

Stouffer Hotels-Inns
(800) 468-3571 USA and Canada

Super 8 Motels, Inc.
(800) 800-8000 USA and Canada

Superior (See Best Value Inns entry)

Susse Chalet
(800) S-CHALET USA and Canada

Travelodge & Viscount Hotels
(800) 255-3050 USA and Canada

Vagabond Inns, Inc.
(800) 522-1555 USA
(800) 468-2251 Canada

WestCoast Hotels, Inc.
(800) 426-0670 USA and Canada

Westin Hotels & Resorts
(800) 228-3000 USA and Canada

Wyndham Hotels & Resorts
(800) 822-4200 Cont'l USA
(800) 631-4200 Canada

Clear Channel Radio Stations

Stations listed are American and Canadian AM unlimited time stations designated to operate with 50 kilowatt power and to render service over an extended area. A clear channel station's signals are generally protected for a distance of up to 750 miles at night.

UNITED STATES

Alaska
650, KYAK, Anchorage
750, KFQD, Anchorage
820, KCBF, Fairbanks

Arizona
1580, KCWW, Tempe
660, KTNN, Window Rock

Arkansas
1090, KAAY, Little Rock

California
940, KFRE, Fresno
640, KFI, Los Angeles
1020, KTNQ, Los Angeles
1070, KNX, Los Angeles
1140, KRAK, Sacramento
1530, KFBK, Sacramento
680, KNBR, San Francisco
740, KCBS, San Francisco
810, KGO, San Francisco
1100, KFAX, San Francisco
1580, KBLA, Santa Monica

Colorado
850, KOA, Denver

Connecticut
1080, WTIC, Hartford

District of Columbia
1500, WTOP, Washington

Florida
710, WAQI, Miami
540, WTGO, Pine Hills

Georgia
750, WSB, Atlanta

Hawaii
870, KAIM, Honolulu

Idaho
670, KBOI, Boise

Illinois
670, WMAQ, Chicago
720, WGN, Chicago
780, WBBM, Chicago
890, WLS, Chicago
1000, WLUP, Chicago

Indiana
1190, WOWO, Fort Wayne

Iowa
1040, WHO, Des Moines
1540, KXEL, Waterloo

Kentucky
840, WHAS, Louisville

Louisiana
870, WWL, New Orleans
1130, KWKH, Shreveport

Maryland
1090, WBAL, Baltimore

Massachusetts
680, WRKO, Boston
850, WHDH, Boston
1030, WBZ, Boston
1510, WSSU, Boston

Michigan
760, WJR, Detroit

Minnesota
830, WCCO, Minneapolis
1500, KSTP, St. Paul

Missouri
1120, KMOX, St. Louis

Nebraska
880, KRVN, Lexington
1110, KFAB, Omaha

Nevada
720, KDWN, Las Vegas
780, KROW, Reno

New Mexico
770, KKOB, Albuquerque
1020, KCKN, Roswell

New York
1540, WPTR, Albany
1520, WWKB, Buffalo
660, WFAN, New York
710, WOR, New York
770, WABC, New York
880, WCBS, New York
1010, WINS, New York
1050, WEVD, New York
1130, WNEW, New York
1560, WQXR, New York
1180, WHAM, Rochester
810, WGY, Schenectady

North Carolina
1110, WBT, Charlotte
680, WPTF, Raleigh

Ohio
700, WLW, Cincinnati
1530, WCKY, Cincinnati
1100, WWWE, Cleveland
1220, WKNR, Cleveland

Oklahoma
1520, KOMA, Oklahoma City
1170, KVOO, Tulsa

Oregon
1120, KPNW, Eugene
1190, KEX, Portland

Pennsylvania
1060, KYW, Philadelphia
1210, WOGL, Philadelphia
1020, KDKA, Pittsburgh

Tennessee
650, WSM, Nashville
1510, WLAC, Nashville

Texas
1080, KRLD, Dallas
820, WBAP, Fort Worth
740, KTRH, Houston
1200, WOAI, San Antonio

Utah
1160, KSL, Salt Lake City

Virginia
1140, WRVA, Richmond

Washington
710, KIRO, Seattle
1000, KOMO, Seattle
1090, KING, Seattle
1510, KGA, Spokane

West Virginia
1170, WWVA, Wheeling

Wyoming
1030, KTWO, Casper

CANADA

Alberta
1010, CBR, Calgary

British Columbia
1130, CKWK, Vancouver

Manitoba
990, CBW, Winnipeg

New Brunswick
1070, CBA, Moncton

Ontario
740, CBL, Toronto
860, CJBC, Toronto

Québec
690, CBF, Montréal
940, CBM, Montréal

Climate USA

State	City	Winter (Dec–Feb)				Spring (Mar–May)				Summer (June–Aug)				Fall (Sept–Nov)			
		Maximum Normal Daily Temp. (F)	Minimum Normal Daily Temp. (F)	Total Precipitation (In)	Total Days with Precipitation	Maximum Normal Daily Temp. (F)	Minimum Normal Daily Temp. (F)	Total Precipitation (In)	Total Days with Precipitation	Maximum Normal Daily Temp. (F)	Minimum Normal Daily Temp. (F)	Total Precipitation (In)	Total Days with Precipitation	Maximum Normal Daily Temp. (F)	Minimum Normal Daily Temp. (F)	Total Precipitation (In)	Total Days with Precipitation
Alabama	Birmingham	56	38	16	32	75	53	15	35	90	70	13	29	76	54	10	23
	Mobile	63	43	15	32	77	57	17	26	90	72	22	42	78	58	12	24
Alaska	Juneau	31	20	12	56	46	31	10	52	63	46	12	50	47	36	19	63
Arizona	Phoenix	67	39	2	12	84	52	1	6	103	74	2	10	87	57	2	8
	Tucson	65	39	2	12	81	50	1	7	97	71	5	21	83	56	2	11
Arkansas	Little Rock	52	31	13	28	71	49	15	30	90	68	10	23	74	49	10	21
California	Los Angeles	67	48	8	16	71	53	3	11	81	62	.07	2	78	58	2	6
	Sacramento	55	39	10	28	72	46	4	17	91	57	.2	2	76	50	3	11
	San Diego	65	47	5	19	68	54	2	14	74	63	.13	1	73	58	2	8
	San Francisco	57	47	12	31	61	49	5	19	64	53	.2	3	67	54	4	13
Colorado	Denver	45	18	2	17	60	34	6	29	84	56	5	27	66	37	3	16
Connecticut	Hartford	35	18	10	33	58	36	11	33	82	59	11	31	63	41	11	29
Delaware	Wilmington	42	25	9	30	62	42	10	34	84	64	11	28	67	47	9	25
District of Columbia	Washington		29	8	28	66	45	10	32	86	67	12	28	69	50	9	23
Florida	Jacksonville	66	45	9	24	79	57	10	24	89	71	21	41	79	61	14	27
	Miami	76	59	6	19	82	67	12	23	89	75	23	48	84	70	20	40
	Tampa	71	51	7	19	82	61	8	19	90	73	23	45	83	65	11	25
Georgia	Atlanta	53	34	13	32	70	50	14	30	86	68	12	31	72	52	9	21
Hawaii	Honolulu	81	66	10	30	83	69	6	28	87	73	2	21	86	71	6	25
Idaho	Boise	40	24	4	34	61	37	3	25	85	55	1	11	64	39	2	20
Illinois	Chicago	33	17	5	32	58	37	9	37	81	59	10	29	63	41	7	29
	Peoria	34	18	5	27	60	40	11	35	84	63	11	27	64	43	8	24
Indiana	Indianapolis	38	22	8	34	61	41	11	37	84	63	11	28	65	44	8	26
	South Bend	33	19	7	43	57	37	10	38	81	60	11	29	62	42	9	31
Iowa	Des Moines	31	15	3	23	58	38	9	31	83	63	11	28	62	42	7	22
Kansas	Wichita	44	24	3	16	67	44	8	26	90	68	12	24	70	47	7	19
Kentucky	Lexington	43	26	11	36	65	44	13	37	85	64	12	31	67	46	8	27
	Louisville	44	26	10	34	65	44	13	36	86	65	11	29	68	46	9	25
Louisiana	New Orleans	64	45	14	29	78	58	14	24	90	72	17	38	79	60	12	22
	Shreveport	59	39	12	27	76	55	14	27	92	72	9	22	78	56	9	21
Maine	Portland	33	13	11	33	52	32	10	35	77	54	8	28	59	38	11	29
Maryland	Baltimore	43	26	9	28	64	42	10	33	85	64	12	28	68	47	9	23
Massachusetts	Boston	37	24	11	35	56	41	11	34	79	62	9	30	62	48	11	29
Michigan	Detroit	34	21	6	38	56	38	8	37	81	61	9	29	62	45	7	29
	Grand Rapids	32	18	6	44	56	35	9	37	81	58	9	29	61	41	9	34
	Sault Ste. Marie	24	8	6	53	47	28	7	35	73	50	10	33	52	36	10	43
Minnesota	Duluth	21	2	3	34	47	27	8	34	73	52	12	35	51	34	7	31
	Minneapolis-St. Paul	24	7	2	25	53	33	7	31	80	59	11	32	57	37	6	25
Mississippi	Jackson	60	37	14	30	77	52	15	29	92	69	11	29	79	52	9	23
Missouri	Kansas City	43	26	4	21	64	45	10	30	89	69	11	26	70	48	8	21
	St. Louis	42	25	6	26	65	45	11	33	87	67	11	26	68	48	8	25
Montana	Billings	35	16	2	22	54	33	5	29	81	55	4	24	59	37	3	19
	Great Falls	33	14	3	24	54	31	5	30	80	51	5	26	58	35	3	20
Nebraska	Lincoln	36	15	3	17	61	39	8	29	86	64	12	25	65	42	6	20
	Omaha	36	16	2	20	62	39	9	30	86	64	13	28	66	42	6	19
Nevada	Las Vegas	58	34	1	8	78	50	1	6	101	72	1	7	80	53	1	6
	Reno	47	21	3	18	64	31	2	15	87	45	1	8	69	31	1	10

*Includes the liquid water equivalent of snow—(10" of snow equals approx. 1" of liquid water).

State	City	Winter (Dec–Feb)				Spring (Mar–May)				Summer (June–Aug)				Fall (Sept–Nov)			
		Maximum Normal Daily Temp. (F)	Minimum Normal Daily Temp. (F)	Total Precipitation (In)	Total Days with Precipitation	Maximum Normal Daily Temp. (F)	Minimum Normal Daily Temp. (F)	Total Precipitation (In)	Total Days with Precipitation	Maximum Normal Daily Temp. (F)	Minimum Normal Daily Temp. (F)	Total Precipitation (In)	Total Days with Precipitation	Maximum Normal Daily Temp. (F)	Minimum Normal Daily Temp. (F)	Total Precipitation (In)	Total Days with Precipitation
New Hampshire	Concord	33	12	8	32	56	32	9	34	80	54	9	31	61	37	10	29
New Jersey	Atlantic City	43	24	10	31	61	50	10	32	82	62	12	31	66	46	10	29
	Newark	40	25	9	32	60	42	11	34	83	65	11	29	66	48	10	26
New Mexico	Albuquerque	49	25	1	12	70	41	1	11	90	63	3	22	71	44	2	14
New York	Albany	32	15	7	35	57	35	8	37	81	58	9	31	61	40	8	31
	Buffalo	31	19	8	57	52	36	9	42	77	59	9	31	59	43	1C	38
	New York	40	27	9	31	60	43	10	34	83	66	11	30	66	50	10	25
	Rochester	33	18	7	52	55	35	8	40	80	58	8	30	61	42	8	37
	Syracuse	33	18	8	54	52	36	9	44	80	59	10	33	61	43	9	39
North Carolina	Charlotte	53	32	11	30	72	48	11	30	87	67	12	31	72	51	9	21
	Greensboro	50	29	10	30	70	46	10	30	86	65	13	34	71	47	9	23
	Raleigh	52	30	10	29	71	46	10	29	87	65	14	30	72	48	9	23
North Dakota	Bismarck	23	2	1	23	52	29	4	26	81	55	8	29	57	32	3	19
Ohio	Cincinnati	41	26	9	34	64	44	12	37	85	64	11	31	67	47	8	28
	Cleveland	35	22	7	46	57	38	10	43	80	59	10	31	62	44	8	35
	Columbus	38	21	7	38	62	39	11	40	83	60	11	31	65	42	7	28
Oklahoma	Oklahoma City	50	28	4	16	70	48	11	25	91	69	9	22	73	50	7	18
	Tulsa	50	29	5	20	70	48	12	28	91	69	11	23	73	50	9	19
Oregon	Portland	46	34	16	54	60	41	8	43	76	54	3	18	63	44	11	39
Pennsylvania	Harrisburg	39	24	8	32	63	41	10	36	85	63	10	29	66	45	8	24
	Philadelphia	42	26	9	30	63	42	10	33	85	64	12	28	67	47	9	25
	Pittsburgh	37	22	8	47	60	39	11	41	81	59	10	33	63	43	7	32
Rhode Island	Providence	38	22	11	33	56	38	11	34	79	60	9	30	63	44	11	27
South Carolina	Columbia	58	34	10	29	76	51	11	28	91	69	15	33	76	52	9	21
	Greenville	53	34	12	30	71	49	12	32	87	68	12	32	72	51	10	25
South Dakota	Rapid City	37	13	1	20	56	31	6	31	83	56	7	29	62	35	2	16
	Sioux Falls	27	7	2	18	56	34	7	27	83	59	15	29	60	36	9	20
Tennessee	Chattanooga	51	31	16	34	72	47	13	32	89	66	22	33	73	48	10	24
	Knoxville	50	33	14	35	71	48	12	35	87	67	11	31	71	50	9	27
	Memphis	51	33	14	29	71	51	15	30	90	70	10	25	73	51	9	21
	Nashville	49	30	14	33	70	48	13	34	89	67	10	28	72	49	9	24
Texas	Corpus Christi	68	48	5	22	81	62	6	17	93	75	8	17	83	63	9	22
	Dallas	58	36	6	19	75	53	11	24	94	73	7	16	78	56	8	19
	El Paso	59	32	1	11	78	49	1	6	94	68	3	18	77	49	2	12
	Houston	65	43	11	26	79	58	11	25	93	72	13	28	82	58	13	24
	San Antonio	64	42	5	23	80	58	7	23	95	73	7	15	81	59	8	19
Utah	Salt Lake City	40	21	4	28	62	36	5	28	88	57	3	15	65	39	3	18
Vermont	Burlington	28	10	6	41	52	32	7	38	79	56	11	37	58	39	9	37
Virginia	Norfolk	50	33	10	29	67	48	9	31	85	68	15	30	70	53	10	24
	Richmond	49	28	9	28	69	45	9	31	87	65	14	31	71	48	10	23
Washington	Seattle-Tacoma	47	36	14	55	59	42	7	41	73	54	3	20	61	46	10	36
	Spokane	35	23	6	43	56	36	4	29	80	53	2	17	57	38	4	26
West Virginia	Charleston	45	26	10	44	66	43	11	42	84	62	12	35	68	45	8	30
	Parkersburg	43	26	9	40	65	43	11	39	84	63	12	33	67	46	7	29
Wisconsin	Madison	28	11	4	28	54	33	8	33	79	57	11	31	58	38	7	27
	Milwaukee	30	14	4	31	53	34	8	36	78	57	10	30	59	40	7	28
Wyoming	Cheyenne	40	17	1	17	55	30	5	30	80	52	6	32	61	34	2	18

*Includes the liquid water equivalent of snowfall–(10" of snow equals approx. 1" of liquid water).

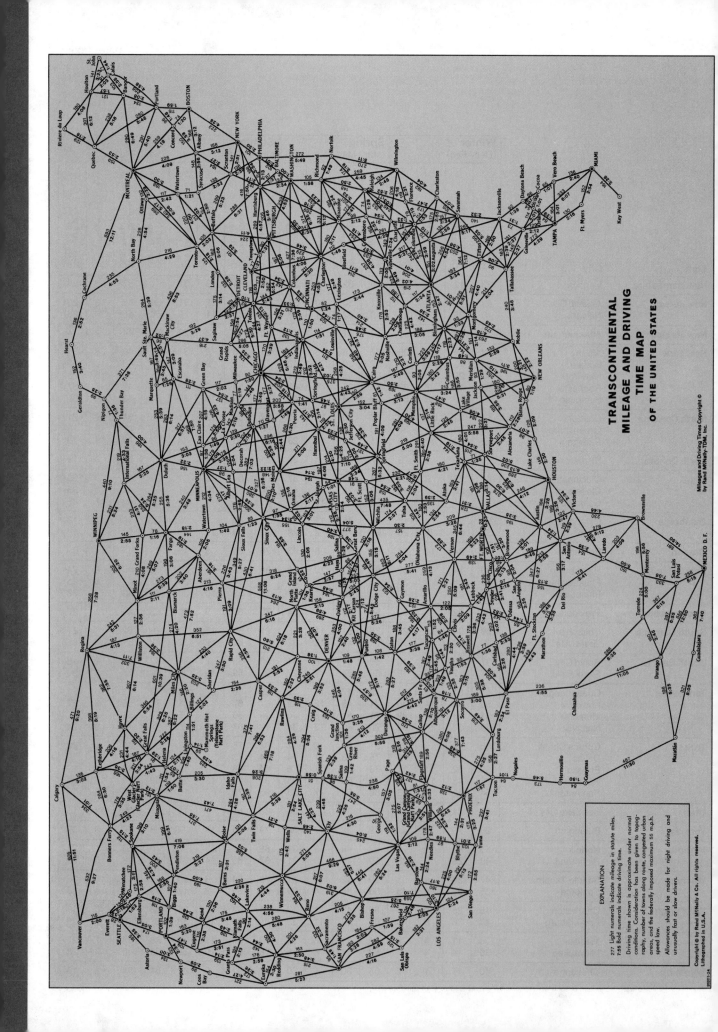

TRANSCONTINENTAL
MILEAGE AND DRIVING
TIME MAP
OF THE UNITED STATES

Mileages and Driving Times Copyright ©
by Rand McNally-TDM, Inc.

Maps: United States, Canada, and Mexico

Introduction

One of the most important parts of any driving trip is knowing the best way to get to your destination. Using the detailed, accurate maps found in this atlas, you can easily plan your trip to avoid the all-too-common frustrations of wrong turns and misjudged mileage.

The maps in this section include a United States map; maps of each of the 50 states; a Canada map; and maps of Alberta, Atlantic provinces, British Columbia, Manitoba, Ontario, Québec, and Saskatchewan, and Mexico.

Among the many features to be found on the maps, the following are of special use in trip planning:

Detailed Transportation Routes. The atlas differentiates between toll roads, freeways, routes under construction, and various secondary roads.

Place Indexes. An index to cities and towns is conveniently located with each state and province map, as well as the map of Mexico. Capitals are noted at the head of each index; land area and total population are also provided.

The United States map includes indexes to National Parks and National Monuments.

Distance Guides. Mileage can be computed by using either the map scale included with each map, or by referring to the accumulated mileage numbers that appear between the red pointers located along major highways.

Trip Planning

A key element to a successful trip is determining the exact location of your destination and the most direct way to get there.

Locating the Destination. To locate your destination, look in the alphabetical listing of cities and towns accompanying the appropriate map. Each town or city has a map reference key, composed of a letter and a number. To find a town on the map with a reference key of B-5, for example, look down the side of the map for the letter B, and draw an imaginary line across the map until it intersects with an imaginery line drawn up or down from the number 5. Your destination will be within an inch-and-a-quarter square surrounding this point.

Marking the Route. Once you have located your destination, determine the most direct route by considering both distance and type of road. A secondary road may seem to provide a shortcut, while in actuality it takes more time because of lower speed limits, unimproved surface conditions, etc.

If time is a consideration, toll roads and freeways most often provide the quickest ride and should be used wherever possible.

To ease your map reading while driving, mark your chosen route with a see-through, felt-tip marker in a color that won't conflict with the lines, names, or symbols shown on the map.

Legend

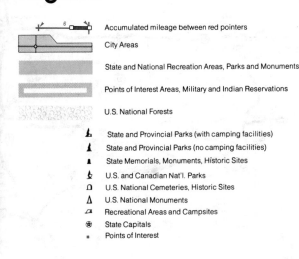

COMPARATIVE DISTANCE SCALE

km. 0 5 10 20 30 40 50 60 70 80 90 100   1 mile = 1.609 kilometers

mi. 0 5 10 20 30 40 50 60   1 kilometer = .621 miles

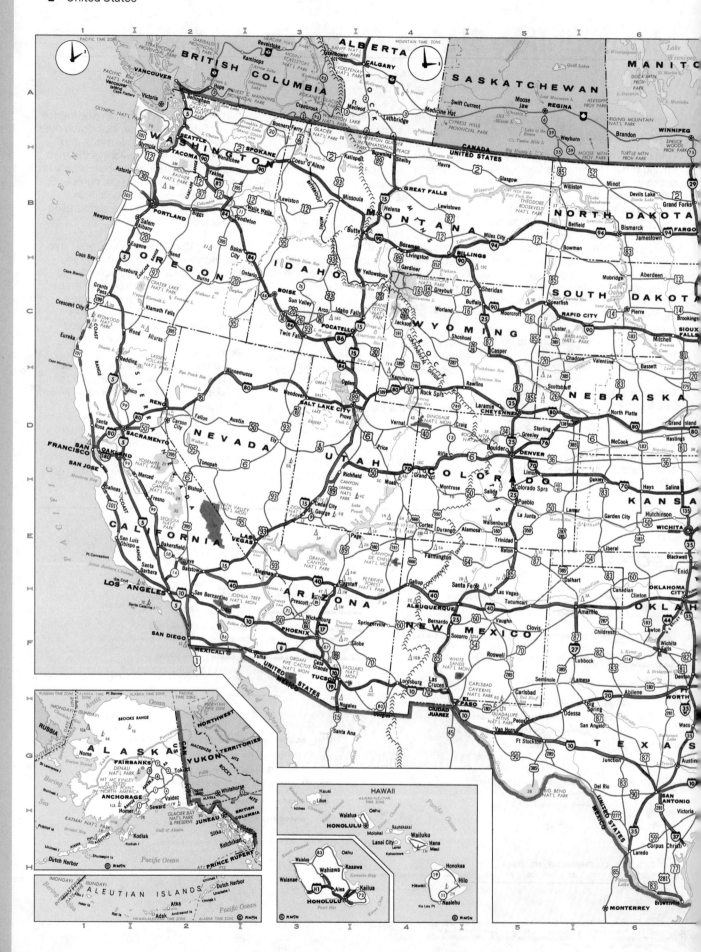

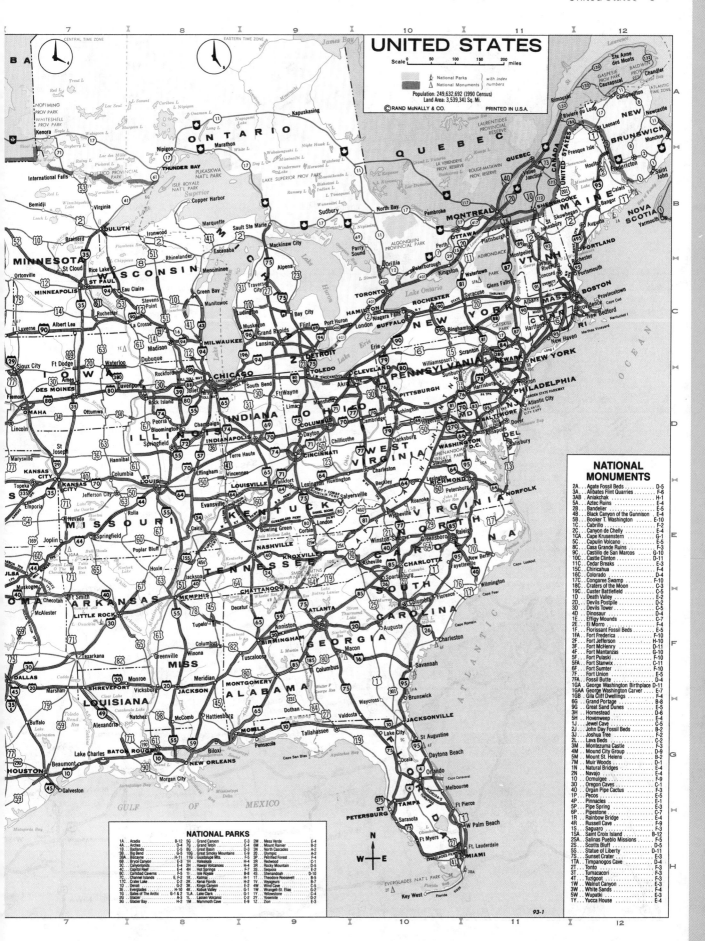

UNITED STATES

Scale 0 50 100 150 200 miles

🏕 National Parks — with index numbers
△ National Monuments

Population: 249,632,692 (1990 Census)
Land Area: 3,539,341 Sq. Mi.
ⓒRAND McNALLY & CO.   PRINTED IN U.S.A.

NATIONAL MONUMENTS

2A	Agate Fossil Beds	D-5
3A	Alibates Flint Quarries	F-6
3AB	Aniakchak	H-1
5A	Aztec Ruins	E-4
2B	Bandelier	E-5
4B	Black Canyon of the Gunnison	E-4
5B	Booker T. Washington	E-10
1C	Cabrillo	F-2
2C	Canyon de Chelly	E-4
1CA	Cape Krusenstern	G-1
5C	Capulin Volcano	E-5
6C	Casa Grande Ruins	F-3
9C	Castillo de San Marcos	G-10
10C	Castle Clinton	D-11
11C	Cedar Breaks	E-3
15C	Chiricahua	F-4
16C	Colorado	D-4
17C	Congaree Swamp	F-10
18C	Craters of the Moon	C-3
19C	Custer Battlefield	C-5
1D	Death Valley	E-2
2D	Devils Postpile	D-2
3D	Devils Tower	C-5
4D	Dinosaur	D-4
1E	Effigy Mounds	D-8
2E	El Morro	F-4
1FA	Florissant Fossil Beds	E-5
1F	Fort Frederica	F-10
2F	Fort Jefferson	H-10
3F	Fort McHenry	D-11
4F	Fort Matanzas	G-10
5F	Fort Pulaski	F-10
5FA	Fort Stanwix	C-11
6F	Fort Sumter	F-10
7F	Fort Union	E-5
7FA	Fossil Butte	D-4
1GA	George Washington Birthplace	D-11
1GAA	George Washington Carver	E-7
1GB	Gila Cliff Dwellings	F-4
6G	Grand Portage	B-8
9G	Great Sand Dunes	E-5
3H	Homestead	D-6
5H	Hovenweep	E-4
1J	Jewel Cave	C-5
2J	John Day Fossil Beds	B-2
3J	Joshua Tree	F-2
2L	Lava Beds	C-2
3M	Montezuma Castle	F-3
4M	Mound City Group	D-9
5M	Mount St. Helens	B-2
7M	Muir Woods	D-1
1N	Natural Bridges	E-4
2N	Navajo	E-4
1O	Ocmulgee	F-9
3O	Oregon Caves	C-1
1P	Pecos	E-5
4P	Pinnacles	D-2
5P	Pipe Spring	E-3
6P	Pipestone	C-7
1R	Rainbow Bridge	E-4
4R	Russell Cave	F-9
1S	Saguaro	F-3
15A	Saint Croix Island	B-12
2SA	Salinas Pueblo Missions	F-5
5S	Scotts Bluff	D-5
5SS	Statue of Liberty	D-11
7S	Sunset Crater	E-3
1TA	Timpanogos Cave	D-4
2T	Tonto	F-3
3T	Tumacacori	F-3
4T	Tuzigoot	F-3
1W	Walnut Canyon	E-3
3W	White Sands	F-4
5W	Wupatki	F-3
1Y	Yucca House	E-4

NATIONAL PARKS

1A	Acadia	B-12	
4A	Arches	D-4	
1B	Badlands	C-5	
3B	Big Bend	G-5	
3BA	Biscayne	H-11	
6B	Bryce Canyon	E-3	
4C	Canyonlands	E-4	
6C	Carlsbad Caverns	F-5	
7C	Channel Islands	E-2	
17C	Crater Lake	C-2	
1D	Denali	G-2	
3E	Everglades	H-10	
1G	Gates Of The Arctic	G-1 & 2	
2G	Glacier	A-3	
3G	Glacier Bay	H-2	
5G	Grand Canyon	E-3	
7G	Grand Teton	D-4	
8G	Great Basin	D-3	
10G	Great Smoky Mountains	E-9	
11G	Guadalupe Mts.	F-5	
1H	Haleakala	H-5	
2H	Hawaii Volcanoes	H-5	
4H	Hot Springs	F-7	
5H	Isle Royale	B-8	
1K	Katmai	H-1	
2K	Kenai Fjords	H-2	
3K	Kings Canyon	D-2	
4K	Kobuk Valley	G-1	
1LA	Lake Clark	G-1	
1L	Lassen Volcanic	C-2	
1M	Mammoth Cave	E-9	
2M	Mesa Verde	E-4	
4M	Mount Rainier	B-2	
3N	North Cascades	A-2	
2O	Olympic	A-2	
3P	Petrified Forest	F-4	
4R	Redwood	C-1	
3R	Rocky Mountain	D-5	
7S	Sequoia	E-2	
8S	Shenandoah	D-10	
1T	Theodore Roosevelt	B-5	
1V	Voyageurs	B-7	
4W	Wind Cave	C-5	
5W	Wrangell-St. Elias	G-2	
1Y	Yellowstone	C-4	
1YA	Yosemite	D-2	
1Z	Zion	E-3	

93-1

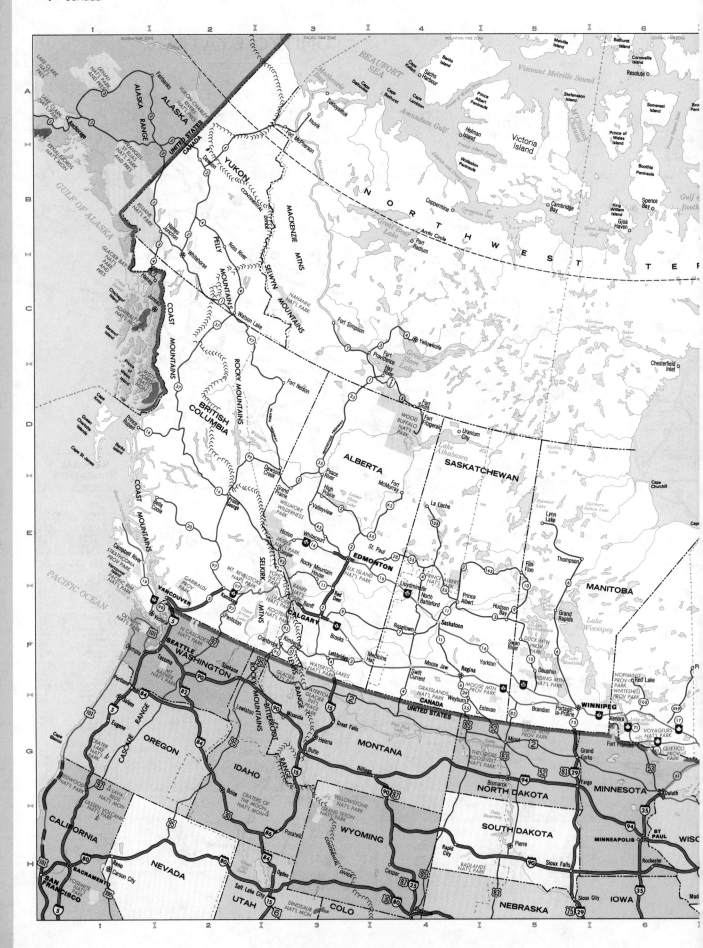

EASTERN TIME ZONE   ATLANTIC TIME ZONE   GREENLAND TIME ZONE

CANADA
Scale 0 50 100 150 200 250 miles
National Parks
National Monuments
Population: 27,296,859 (1991 Census)
Land Area: 3,851,809 Sq. Mi.
©RAND McNALLY & CO.   PRINTED IN U.S.A.

Devon Island
Lancaster Sound
Cape Liverpool
Bylot Island
Borden Peninsula
Arctic Bay
Igloolik
BAFFIN BAY
GREENLAND
Godthab
Baffin Island
Prince Charles Island
AUYUITTUQ NAT'L PARK
Pangnirtung
DAVIS STRAIT
Melville Peninsula
Foxe Basin
Arctic Circle
Nettilling Lake
Amadjuak Lake
Iqaluit
Hall Peninsula
Southampton Island
Nottingham Island
Hudson Strait
Resolution Island
Ivujivik
Cape Chidley
LABRADOR SEA
Fisher Strait
Cape Kendall
Cape Low
Coats
Cape Southampton
Mansel Island
Akpatok Island
Ungava Bay
Hebron
TERRITORIES
HUDSON BAY
Belcher Islands
Ottawa Islands
Lac d' Iberville
Fort Chimo
Cape Harrison
Battle Harbour
St. Anthony
NEWFOUNDLAND
Goose Bay
Lake Melville
Tatnam
Cape Henrietta Maria
Winisk
POLAR BEAR PROV PARK
James Bay
Akimiski Island
Charlton I
Mushalagan Lake
QUEBEC
Lac d' Iberville
Labrador City
Ashuanipi Lake
Cape Freels
GROS MORNE NAT'L PARK
Bonavista
St. John's
TERRA NOVA NAT'L PARK
Cape Race
Grand Bank
Corner Brook
Channel-Port-aux-Basques
St. Pierre and Miquelon (France)
ONTARIO
Kettle Lake
Lake Nipigon
Waskaganish
Lac Mistassini
MISTASSINI PROV RES
Chibougamau
Reservoir Pipmuacanau
Baie-Comeau
Havre-St. Pierre
Sept-Iles
Ile Anticosti
Detroit de Jacques · Cartier
Detroit d' Honguedo
Gulf of St. Lawrence
Cape Breton Island
Sable Island
Hearst
Geraldton
Nipigon
Matagami
Chicoutimi
Rimouski
Campbellton
Bathurst
KOUCHIBOUGUAC NAT'L PARK
PRINCE EDWARD NAT'L PARK
CAPE BRETON HIGHLANDS NAT'L PARK
Sydney
GASPESIE PROV PARK
PEI
Charlottetown
HAUTE MAURICE PROV PARK
Rouyn-Noranda
Val-d'-Or
LAURENTIDES PROV RES
Riviere-du-Loup
Edmundston
NEW BRUNSWICK
Moncton
NOVA SCOTIA
Thunder Bay
PUKASKWA NAT'L PARK
ISLE ROYALE NAT'L PARK
Kewenaw Point
Lake Superior
LAKE SUPERIOR PROV PARK
Wawa
Timmins
LA VERENDRYE PROV RES
ROUGE-MATAWIN PROV RES
La Tuque
LA MAURICE NAT'L PARK
Trois-Rivieres
MAINE
APPALACHIAN MTNS
FUNDY NAT'L PARK
Saint John
KEJIMKUJIK NAT'L PARK
Dartmouth
Halifax
Sault Ste Marie
Sudbury
North Bay
Pembroke
Mont-Laurier
MONTREAL
Quebec
Drummondville
Sherbrooke
Bangor
ACADIA NAT'L PARK
Cape Sable
ATLANTIC OCEAN
Mackinaw City
MICHIGAN
GEORGIAN BAY IS NAT'L PARK
ALGONQUIN PROV PARK
Peterborough
Hull
OTTAWA
Kingston
Watertown
LAWRENCE Montpelier
VT
Augusta
PORTLAND
Green Bay
Lake Huron
Georgian Bay
Manitoulin I
TORONTO
HAMILTON
Oshawa
Lake Ontario
Rochester
Syracuse
Albany
NH
Concord
MASS
BOSTON
Cape Cod
RI
Providence
MILWAUKEE
Grand Rapids
Lansing
LONDON
Kitchener
Niagara Falls
BUFFALO
NEW YORK
Syracuse
HARTFORD
CONN
Lake Michigan
Sarnia
Lake Erie
PT PELEE NAT'L PARK
Windsor
DETROIT
CHICAGO
ILL
PENNSYLVANIA
Scranton
NEW YORK
Long Island

93-1

ALABAMA

Population: 4,062,608
(1990 Census)
Land Area: 50,766 Sq. Mi.
Capital: Montgomery

Cities and Towns

Abbeville F-5
Alabaster D-3
Albertville B-4
Alexander City D-4
Aliceville D-1
Andalusia G-3
Anniston C-4
Ashford G-5
Ashland C-4
Ashville C-4
Athens A-3
Atmore G-2
Attalla B-4
Auburn E-5
Bay Minette G-2
Bellamy E-1
Bessemer C-3
Birmingham C-3
Boaz B-4
Brewton G-3
Bridgeport A-4
Brundidge F-4
Butler E-1
Camden F-2
Castleberry G-3
Centre B-4
Centreville D-2
Chatom F-1
Childersburg C-3
Clanton D-3
Clayton F-5
Cleveland B-3
Collinsville B-4
Columbiana C-3
Cottondale D-2
Courtland A-2
Cullman B-3
Dadeville D-4
Daleville G-4
Decatur A-3
Demopolis E-2
Dothan G-5
Double Springs B-2
E. Brewton G-3
Elba F-4
Enterprise G-4
Eufaula F-5
Eutaw D-1
Evergreen F-3
Fairfield C-3
Fairhope H-1
Fayette C-2
Florala G-4
Florence A-2
Ft. Deposit F-3
Ft. Payne B-4
Frisco City F-2
Gadsden B-4
Geneva G-4
Georgiana F-3
Goodwater D-4
Grand Bay H-1
Greensboro D-2
Greenville F-3
Grove Hill F-2
Guntersville B-3
Hackleburg B-2
Haleyville B-2
Hamilton B-1
Hanceville B-3
Hartford G-4
Hartselle B-3
Hayneville E-3
Heflin C-4
Homewood C-3
Huntsville A-3
Jackson F-1
Jacksonville C-4
Jasper C-3
Lafayette D-5
Lanett D-5
Leeds C-3
Linden E-2
Livingston E-1
Loxley H-2
Luverne F-4
Mc Kenzie F-3
Maplesville D-3
Marion E-2
Mobile H-1
Montevallo D-3
Montgomery E-4
Moulton B-2
Mountain Brook C-3
Mt. Vernon G-1
Muscle Shoals A-2
New Brockton F-4
Normal A-3
Northport D-2
Oneonta B-3
Opelika D-5
Opp G-4
Ozark G-4
Phenix City E-5
Piedmont B-4
Prattville E-3
Prichard H-1
Rainbow City B-4
Red Bay B-1
Roanoke D-5
Robertsdale H-2
Rogersville A-2
Russellville B-2
Samson G-4
Saraland G-1
Scottsboro A-4
Selma E-3
Sheffield A-2
Stapleton H-2
Stevenson A-4
Sulligent B-1
Sylacauga D-4
Talladega C-4
Tallassee E-4
Tarrant City C-3
Thomaston E-2
Thomasville F-2
Troy F-4
Tuscaloosa D-2
Tuscumbia A-2
Tuskegee E-4
Union Sprs. E-4
Uniontown E-2
Valley D-5
Vernon C-1
Vincent C-3
Wedowee C-5
Wetumpka E-4
Winfield B-2
York E-1

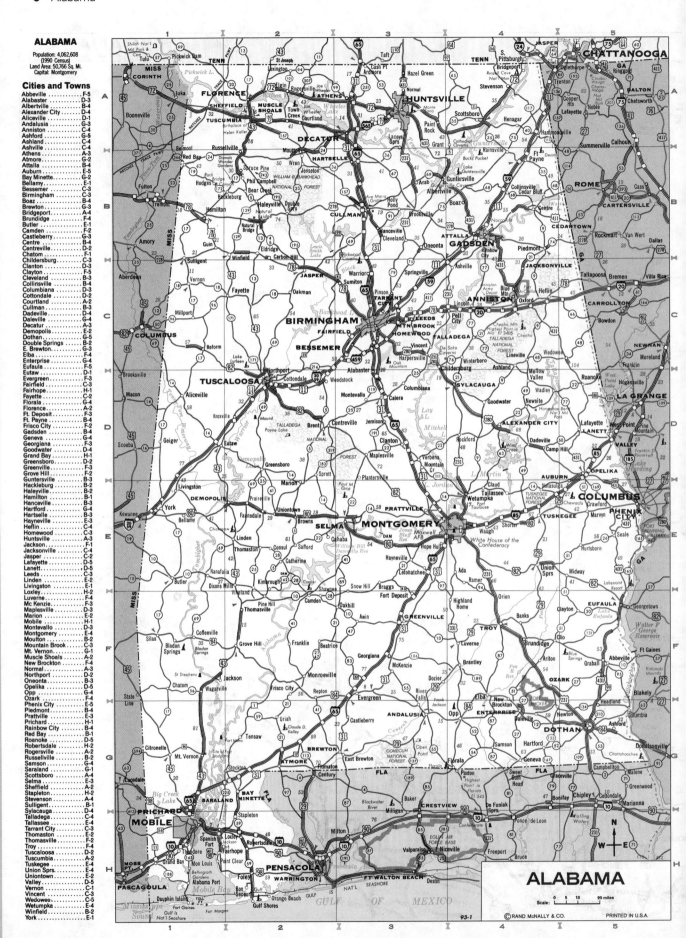

ALABAMA

Scale: 0   5   10   20 miles

93-1   © RAND McNALLY & CO.   PRINTED IN U.S.A.

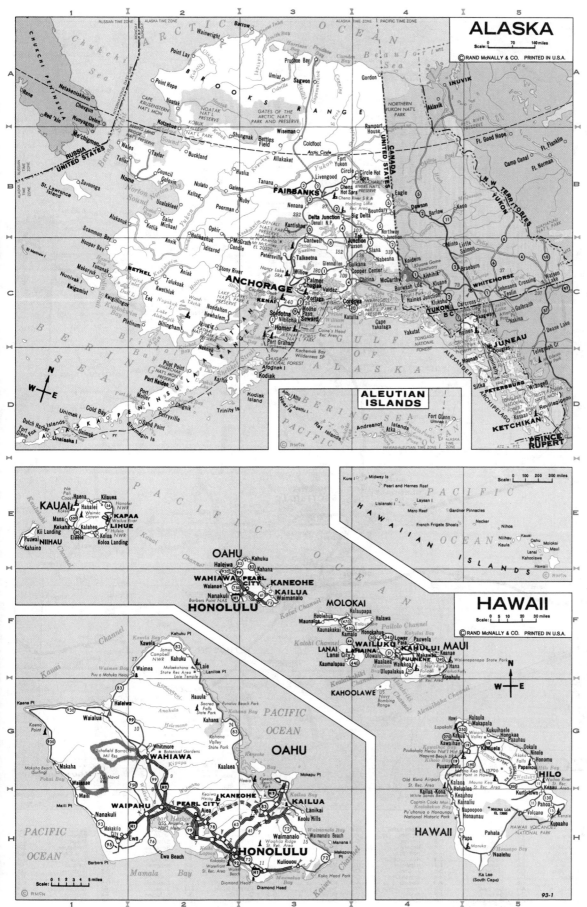

ALASKA

Population: 551,947
(1990 Census)
Land Area: 570,833 Sq. Mi.
Capital: Juneau

Cities and Towns

Alakanuk	B-1
Allakaket	B-3
Anchorage	C-3
Anvik	C-2
Barrow	A-3
Bethel	C-2
Big Delta	B-3
Cantwell	C-3
Chena Hot Sprs.	B-3
Chignik	D-2
Chugiak	C-3
Circle	B-4
Circle Hot Sprs.	B-4
Cold Bay	D-1
Cordova	C-3
Delta Junction	B-3
Dillingham	C-2
Douglas	D-5
Eek	C-2
Fairbanks	B-3
Fort Yukon	B-3
Homer	C-3
Hoonah	D-5
Hooper Bay	C-1
Iditarod	C-2
Inuvik	A-4
Juneau	D-5
Kaltag	B-2
Karluk	D-2
Kenai	C-3
Ketchikan	D-5
Kodiak	D-3
Kotzebue	B-2
Kwethluk	C-2
Kwigillingok	C-1
Livengood	B-3
McGrath	C-3
Nenana	B-3
Newhalem	C-2
Noatak	A-2
Nome	B-2
Nondalton	C-2
Palmer	C-3
Perryville	D-2
Petersburg	D-5
Port Graham	C-3
Pt. Hope	A-2
Prudhoe Bay	A-3
Ruby	B-2
Saint Michael	B-2
Savoonga	B-1
Scammon Bay	C-1
Seward	C-3
Shungnak	B-2
Sitka	D-5
Skagway	C-5
Soldotna	C-3
Spenard	C-3
Tanana	B-3
Taylor	B-2
Tok Junction	B-4
Umiat	A-3
Valdez	C-3
Wainwright	A-2
Willow	C-3
Wrangell	D-5

HAWAII

Population: 1,115,274
(1990 Census)
Land Area: 6,427 Sq. Mi.
Capital: Honolulu

Cities and Towns

Aiea	H-2
Ewa	H-2
Ewa Beach	H-2
Haiaula	H-3
Haleiwa	G-1
Hana	F-4
Hilo	G-5
Holualoa	H-4
Honaunau	H-5
Honokaa	G-5
Honolulu	H-3
Honomu	G-5
Hoolehua	F-3
Kahuku	F-2
Kahului	F-4
Kailua	H-3
Kailua-Kona	H-4
Kainaliu	H-4
Kalaheo	E-1
Kalaupapa	F-3
Kamuela	G-5
Kaneohe	H-3
Kapaa	E-1
Kaunakakai	F-3
Kawaihae	G-4
Keaau	H-5
Kekaha	E-1
Kilauea	E-1
Koloa Landing	E-1
Kukuihaele	G-5
Kurtistown	H-5
Lahaina	F-4
Laie	F-2
Lanai City	F-3
Lihue	E-1
Lower Paia	F-4
Maalaea	F-4
Maili	H-1
Makakilo City	H-1
Makaha	H-1
Makapala	G-5
Makawao	F-4
Maunaloa	F-3
Naalehu	H-5
Nanakuli	H-1
Ninole	G-5
Ookala	G-5
Pahala	H-5
Pahoa	H-5
Papaikou	G-5
Pauwela	F-4
Pearl City	H-2
Puunene	F-4
Volcano	H-5
Wahiawa	G-2
Waialua	G-1
Waianae	G-1
Wailuku	F-4
Waimanalo	H-3
Waimanalo Beach	H-3
Waimea	F-2
Waipahu	H-2
Whitmore	G-2

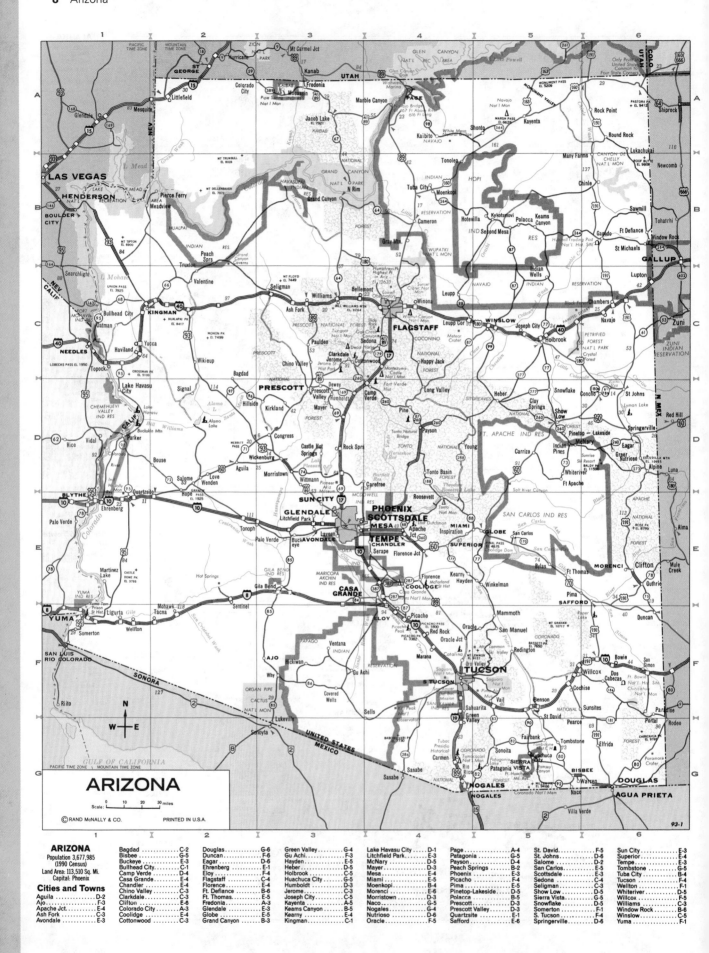

ARIZONA

Population 3,677,985
(1990 Census)
Land Area: 113,510 Sq. Mi.
Capital: Phoenix

Scale: 0 — 10 — 20 — 30 miles

©RAND McNALLY & CO.   PRINTED IN U.S.A.

93-1

Cities and Towns

ARKANSAS

Scale:
0 10 20 30 miles

© RAND McNALLY & CO.

PRINTED IN U.S.A.

93-1

ARKANSAS
Population: 2,362,239
(1990 Census)
Area: 53,104 Sq. Miles
Capital: Little Rock

Cities and Towns

Arkadelphia	E-3
Arkansas City	E-5
Ashdown	E-1
Augusta	C-5
Bald Knob	C-5
Batesville	B-5
Beebe	D-3
Benton	D-3
Bentonville	A-1
Berryville	A-2
Blytheville	B-7
Booneville	C-2
Brinkley	D-5
Bull Shoals	A-3
Camden	E-3
Charleston	C-2
Clarendon	D-5
Clarksville	C-2
Clinton	B-4
Conway	C-4
Corning	A-6
Crossett	E-4
Danville	C-3
Dardanelle	C-3
Des Arc	C-5
De Queen	E-1
De Valls Bluff	D-5
De Witt	D-5
El Dorado	E-4
Eureka Sprs.	A-2
Fayetteville	B-1
Forrest City	C-6
Fordyce	E-4
Ft. Smith	C-1
Glenwood	D-2
Greenwood	C-1
Hamburg	E-4
Hampton	E-3
Harrisburg	B-6
Harrison	A-3
Heber Sprs.	C-4
Helena	D-6
Hope	E-2
Hot Springs Nat'l Pk.	D-2
Hot Springs Village	D-2
Huntsville	B-2
Jacksonville	C-4
Jasper	A-3
Jonesboro	B-6
Lake City	B-6
Lake Village	E-5
Lewisville	E-2
Little Rock	D-4
Lonoke	D-4
McGehee	E-5
Magnolia	E-2
Malvern	D-3
Mammoth Spring	A-5
Marianna	C-6
Marion	B-7
Marked Tree	B-6
Marshall	B-4
Melbourne	B-4
Mena	D-1
Monticello	E-4
Morrilton	C-3
Mountain Home	A-4
Mountain View	B-4
Mt. Ida	D-2
Murfreesboro	E-2
Nashville	E-2
Newport	B-5
N. Little Rock	D-4
Osceola	B-7
Ozark	C-2
Paragould	A-6
Paris	C-2
Perryville	C-3
Piggott	A-6
Pine Bluff	D-4
Pocahontas	A-5
Prescott	E-2
Rison	E-4
Rogers	A-1
Russellville	C-3
Salem	A-4
Searcy	C-4
Sheridan	D-4
Siloam Springs	A-1
Springdale	B-1
Star City	E-4
Stuttgart	D-5
Texarkana	E-1
Trumann	B-6
Van Buren	C-1
Walnut Ridge	B-5
Warren	E-4
W. Helena	D-6
W. Memphis	C-7
Wynne	C-6
Yellville	A-3

CALIFORNIA

Population: 29,839,250
(1990 Census)
Land Area: 156,297 Sq. Mi.
Capital: Sacramento

Cities and Towns

Adelanto	J-5
Alameda	B-4
Alhambra	J-4
Anaheim	J-4
Anderson	E-2
Antioch	E-2
Arvin	I-4
Atascadero	H-2
Atwater	F-3
Auburn	E-3
Avenal	H-3
Avenal	H-3
Bakersfield	I-4
Banning	K-5
Barstow	J-5
Beaumont	K-5
Benicia	E-2
Berkeley	E-1
Beverly Hills	J-4
Bishop	G-5
Blythe	K-7
Bradley	H-3
Brawley	L-6
Brentwood	E-2
Buena Park	J-4
Burbank	J-4
Burney	D-2
Calexico	L-6
Calipatria	L-6
Calistoga	D-2
Cardiff-by-the-Sea	K-4
Carlsbad	K-4
Carpinteria	I-3
Carson	J-4
Chico	E-2
China Lake	H-5
Chowchilla	G-3
Chula Vista	L-5
Claremont	J-4
Cloverdale	D-2
Clovis	G-3
Coalinga	G-3
Colusa	E-2
Concord	E-2
Corcoran	H-4
Corning	E-2
Corona	K-4
Coronado	L-5
Crescent City	A-1
Davis	E-2
Delano	H-4
Desert Hot Spgs.	K-5
Dixon	E-2
Dunsmuir	D-2
Edwards	J-5
El Cajon	L-6
El Centro	L-6
El Monte	J-4
Encinitas	K-4
Escondido	K-5
Eureka	B-1
Exeter	H-4
Fairfield	E-2
Fallbrook	K-5
Fillmore	J-4
Firebaugh	G-3
Fort Bragg	C-1
Fortuna	B-1
Fremont	F-2
Fresno	G-3
Fullerton	J-4
Gilroy	F-2
Glendale	J-4
Grass Valley	E-3
Hanford	H-4
Healdsburg	D-2
Hemet	K-5
Hollister	F-2
Huntington Beach	J-4
Imperial	L-6
Imperial Beach	L-5
Indio	K-6
Kerman	G-3
King City	G-2
La Cañada Flintridge	J-4
Laguna Beach	K-4
La Habra	J-4
Lake Elsinore	K-5
Lakeport	D-2
Lancaster	J-4
Lemoore	H-4
Lincoln	E-3
Lindsay	H-4
Lodi	E-2
Lompoc	I-2
Lone Pine	G-5
Long Beach	J-4
Los Angeles	J-4
Los Banos	G-3

CALIFORNIA

© Rand McNally & Co.

PRINTED IN U.S.A.

93-1

Scale: 0 5 10 20 miles

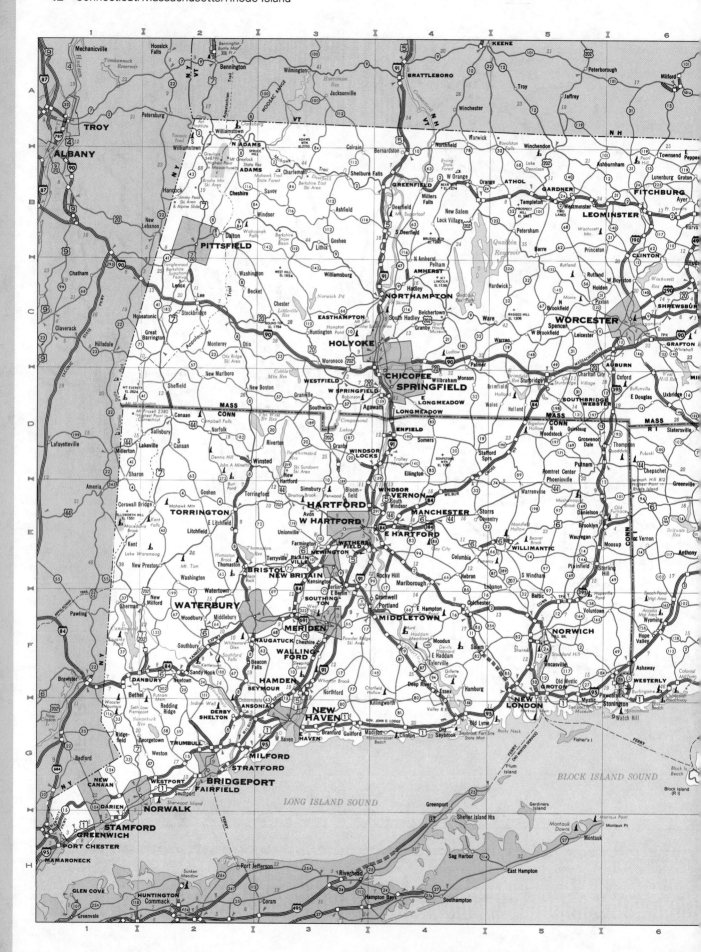

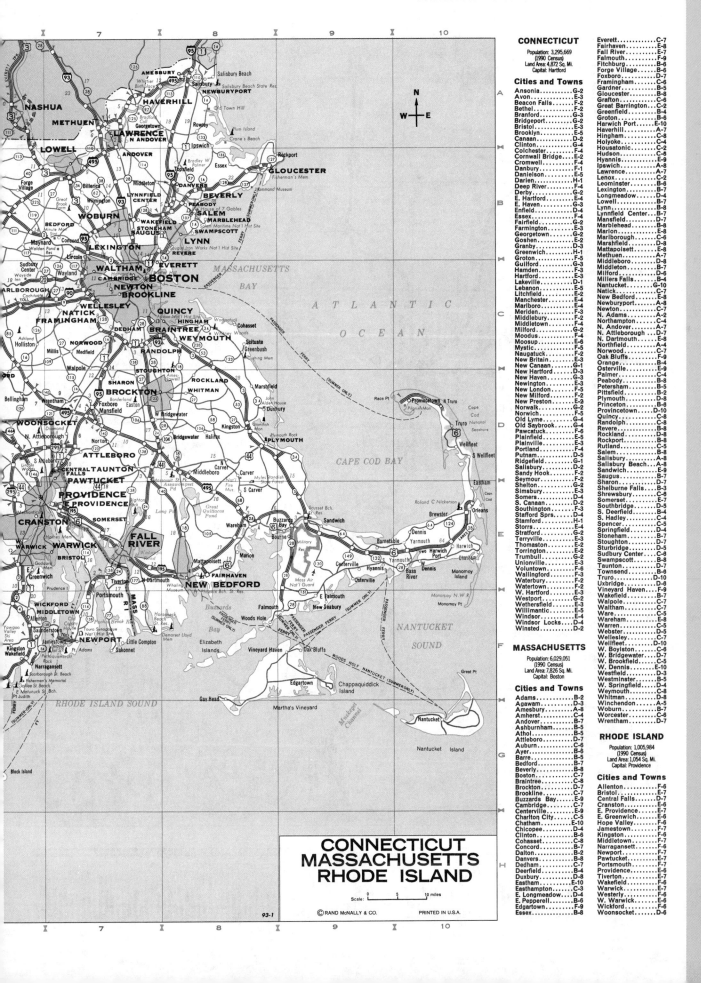

CONNECTICUT

Population: 3,295,669
(1990 Census)
Land Area: 4,872 Sq. Mi.
Capital: Hartford

Cities and Towns

Ansonia............G-2
Avon............E-3
Beacon Falls............F-2
Bethel............F-2
Branford............G-3
Bridgeport............G-2
Bristol............F-2
Brooklyn............D-5
Canaan............D-2
Clinton............G-4
Colchester............F-4
Cornwall Bridge............E-2
Cromwell............F-4
Danbury............F-1
Danielson............D-5
Darien............H-1
Deep River............F-4
Derby............G-2
E. Hartford............E-4
E. Haven............G-3
Enfield............D-4
Essex............F-4
Fairfield............G-2
Farmington............E-3
Georgetown............G-2
Goshen............E-2
Granby............D-3
Greenwich............H-1
Groton............F-5
Guilford............G-3
Hamden............F-3
Hartford............E-4
Lakeville............D-1
Lebanon............E-5
Litchfield............E-2
Manchester............E-4
Marlboro............F-4
Meriden............F-3
Middlebury............F-2
Middletown............F-4
Milford............G-2
Moodus............F-4
Moosup............E-6
Mystic............F-5
Naugatuck............F-2
New Britain............F-3
New Canaan............G-1
New Hartford............D-3
New Haven............G-3
Newington............E-3
New London............F-5
New Milford............F-2
New Preston............E-2
Norwalk............G-2
Norwich............F-5
Old Lyme............G-4
Old Saybrook............G-4
Pawcatuck............F-6
Plainfield............E-5
Plainville............E-3
Portland............F-4
Putnam............D-5
Ridgefield............G-1
Salisbury............D-2
Sandy Hook............F-2
Seymour............F-2
Shelton............G-2
Simsbury............D-3
Somers............D-4
S. Canaan............D-2
Southington............F-3
Stafford Sprs.............D-4
Stamford............H-1
Storrs............E-4
Stratford............G-2
Terryville............E-3
Thomaston............E-2
Torrington............E-2
Trumbull............G-2
Unionville............E-3
Voluntown............F-6
Wallingford............F-3
Waterbury............F-2
Watertown............F-2
W. Hartford............E-3
Westport............G-2
Wethersfield............E-3
Willimantic............E-5
Windsor............E-4
Windsor Locks............D-4
Winsted............D-2

MASSACHUSETTS

Population: 6,029,051
(1990 Census)
Land Area: 7,826 Sq. Mi.
Capital: Boston

Cities and Towns

Adams............B-2
Agawam............D-3
Amesbury............A-8
Amherst............B-7
Andover............A-7
Ashburnham............B-5
Athol............B-5
Attleboro............D-7
Auburn............C-6
Ayer............B-6
Barre............B-5
Bedford............B-7
Beverly............B-8
Boston............C-7
Braintree............D-7
Brockton............D-7
Brookline............C-7
Buzzards Bay............E-9
Cambridge............C-7
Centerville............E-9
Charlton City............C-5
Chatham............E-10
Chicopee............D-4
Clinton............B-6
Cohasset............C-8
Concord............B-7
Dalton............B-2
Danvers............B-8
Dedham............C-7
Deerfield............B-4
Duxbury............D-8
Eastham............E-10
Easthampton............C-3
E. Longmeadow............D-4
E. Pepperell............B-6
Edgartown............F-9
Essex............B-8
Everett............C-7
Fairhaven............E-8
Fall River............E-7
Falmouth............F-9
Fitchburg............B-6
Forge Village............B-6
Foxboro............D-7
Framingham............C-6
Gardner............B-5
Gloucester............B-8
Grafton............C-6
Great Barrington............C-2
Greenfield............B-4
Groton............B-6
Harwich Port............E-10
Haverhill............A-7
Hingham............C-8
Holyoke............C-4
Housatonic............C-2
Hudson............C-6
Hyannis............E-9
Ipswich............A-8
Lawrence............A-7
Lenox............C-2
Leominster............B-6
Lexington............B-7
Longmeadow............D-4
Lowell............B-7
Lynn............B-8
Lynnfield Center............B-7
Mansfield............D-7
Marblehead............B-8
Marion............E-8
Marlborough............C-6
Marshfield............D-8
Mattapoisett............E-8
Methuen............A-7
Middleboro............D-8
Middleton............B-7
Milford............D-6
Millers Falls............B-4
Nantucket............G-10
Natick............C-7
New Bedford............E-8
Newburyport............A-8
Newton............C-7
N. Adams............A-2
Northampton............C-4
N. Andover............A-7
N. Attleborough............D-7
N. Dartmouth............E-8
Northfield............A-4
Norwood............C-7
Oak Bluffs............F-9
Orange............B-4
Osterville............E-9
Palmer............C-4
Peabody............B-8
Petersham............B-5
Pittsfield............B-2
Plymouth............D-8
Princeton............B-5
Provincetown............D-10
Quincy............C-8
Randolph............C-8
Revere............B-8
Rockland............D-8
Rockport............B-8
Rutland............C-5
Salem............B-8
Salisbury............A-8
Salisbury Beach............A-8
Sandwich............E-9
Saugus............B-7
Sharon............D-7
Shelburne Falls............B-3
Shrewsbury............C-6
Somerset............E-7
Southbridge............D-5
S. Deerfield............B-4
S. Hadley............C-4
Spencer............C-5
Springfield............D-4
Stoneham............B-7
Stoughton............D-7
Sturbridge............D-5
Sudbury Center............C-6
Swampscott............B-8
Taunton............D-7
Townsend............B-6
Truro............D-10
Uxbridge............D-6
Vineyard Haven............F-9
Wakefield............B-7
Walpole............C-7
Waltham............C-7
Ware............C-4
Wareham............E-8
Warren............C-5
Webster............D-6
Wellesley............C-7
Wellfleet............D-10
W. Boylston............C-6
W. Bridgewater............D-7
W. Brookfield............C-5
W. Dennis............E-10
Westfield............D-3
Westminster............B-5
W. Springfield............D-4
Weymouth............C-8
Whitman............D-8
Winchendon............A-5
Woburn............B-7
Worcester............C-6
Wrentham............D-7

RHODE ISLAND

Population: 1,005,984
(1990 Census)
Land Area: 1,054 Sq. Mi.
Capital: Providence

Cities and Towns

Allenton............F-6
Bristol............E-7
Central Falls............D-7
Cranston............E-6
E. Providence............E-7
E. Greenwich............E-6
Hope Valley............F-6
Jamestown............F-7
Kingston............F-6
Middletown............F-7
Narragansett............F-6
Newport............F-7
Pawtucket............E-7
Portsmouth............F-7
Providence............E-6
Tiverton............E-7
Wakefield............F-6
Warwick............E-7
Westerly............F-6
W. Warwick............E-6
Wickford............F-6
Woonsocket............D-6

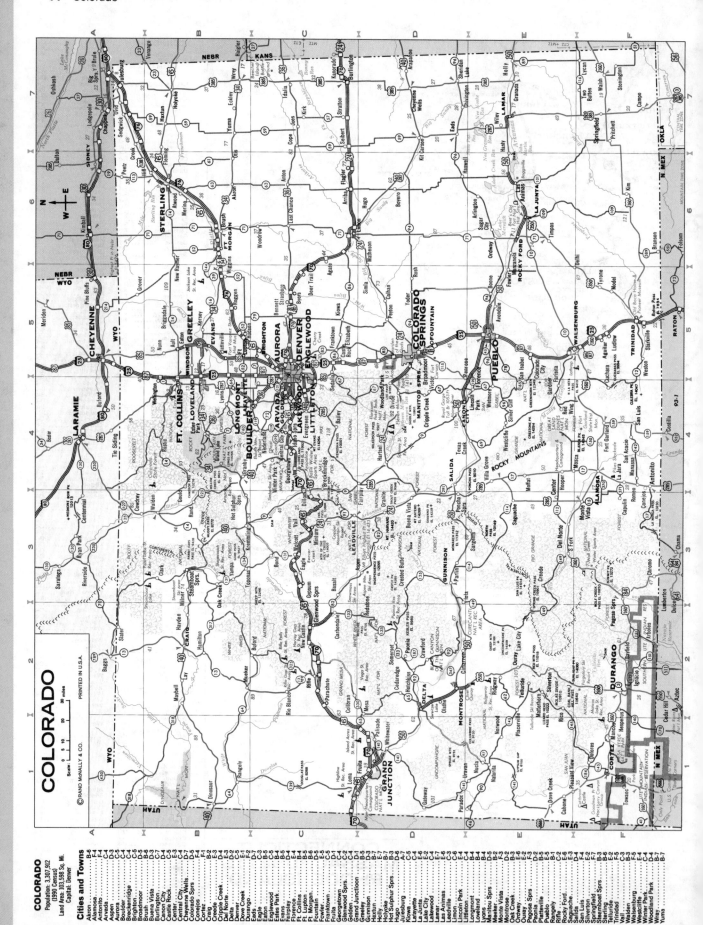

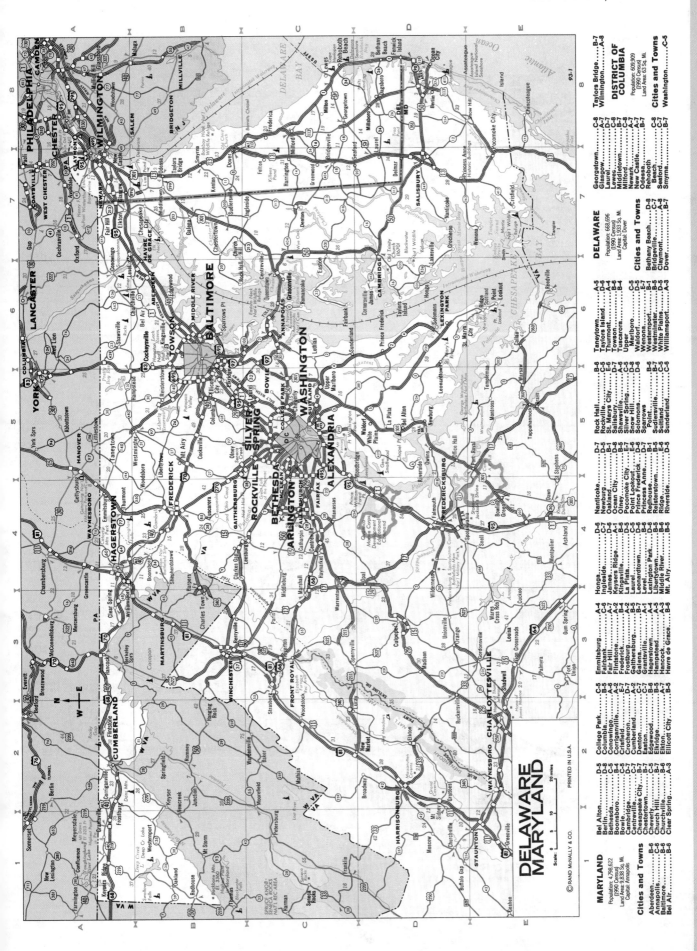

FLORIDA

Population: 13,003,362
(1990 Census)
Land Area: 54,157 Sq. Mi.
Capital: Tallahassee

Cities and Towns

Alachua B-3
Apalachicola H-3
Apopka C-4
Arcadia E-3
Atlantic Beach B-4
Auburndale D-4
Avon Park E-4
Bartow D-4
Belle Glade F-5
Blountstown H-3
Boca Raton F-5
Bonifay G-3
Boynton Beach F-5
Bradenton E-3
Bristol H-3
Bronson C-3
Brooksville D-3
Bunnell C-4
Bushnell C-3
Cantonment G-1
Cape Canaveral ... D-5
Cape Coral F-3
Chattahoochee A-1
Chipley G-3
Clearwater D-3
Clermont D-4
Cocoa D-5
Coral Gables G-5
Crawfordville B-1
Crescent City C-4
Crestview G-2
Cross City B-2
Crystal River C-3
Dade City D-3
Dania F-5
Daytona Beach ... C-4
Deerfield Beach ... F-5
De Funiak Sprs. ... G-2
De Land C-4
Delray Beach F-5
Dunedin D-3
E. Naples F-4
Florida City G-5
Ft. Lauderdale ... F-5
Fort Meade E-3
Ft. Myers E-3
Ft. Ogden E-3
Ft. Pierce E-5
Ft. Walton Beach .. H-2
Frostproof E-4
Gainesville B-3
Goulds G-5
Green Cove Sprs. .. B-4
Greenville A-2
Haines City D-4
Hallandale G-5
Hialeah G-5
High Springs B-3
Holly Hill C-4
Hollywood G-5
Homestead G-5
Inverness D-3
Jacksonville B-4
Jacksonville Beach .. B-4
Jasper A-2
Kenansville D-4
Key West H-4
Kissimmee D-4
La Belle F-4
Lake Butler B-3
Lake City B-3
Lakeland D-3
Lake Placid E-4
Lake Wales E-4
Lake Worth F-5
Lawtey B-3
Leesburg C-3
Live Oak B-2
Madison A-2
Marathon H-4
Marianna H-3
Mayo B-2
Melbourne D-5
Mexico Beach ... H-3
Miami G-5
Miami Beach G-5
Miami Shores G-5
Milton G-1
Monticello A-2
Moore Haven F-4
Naples F-4
New Port Richey .. D-3
New Smyrna Beach . C-4
Ocala C-3
Ochopee G-4
Okeechobee E-4
Opa-Locka G-5
Orange City C-4
Orlando C-4
Ormond Beach ... B-4
Palatka B-4
Palm Beach F-5
Palmetto E-3
Panama City H-3
Pensacola H-1
Perrine G-5
Perry B-2
Plant City D-3
Pompano Beach .. F-5
Punta Gorda F-3
Quincy A-1
Riviera Beach F-5
St. Augustine B-4
St. Cloud D-4
St. Petersburg ... E-3
Sanford C-4
Sarasota E-3
Sebring E-4
Starke B-3
Stuart E-5
Tallahassee A-1
Tampa D-3
Tarpon Sprs. D-3
Tavares C-4
Titusville D-4
Trenton B-3
Venice E-3
Vero Beach E-5
Warrington H-1
Watertown B-3
Wauchula E-3
Weston F-5
W. Palm Beach ... F-5
W. Pensacola ... H-1
Wewahitchka H-3
Winter Haven D-4
Winter Park D-4

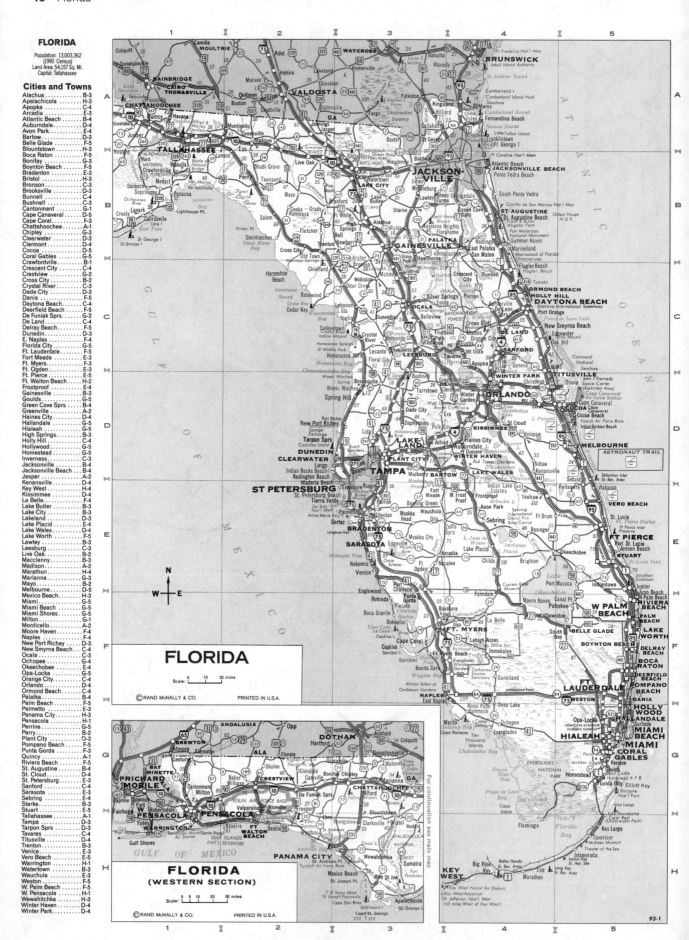

FLORIDA

Scale: 0 10 20 miles

© RAND McNALLY & CO.   PRINTED IN U.S.A.

FLORIDA
(WESTERN SECTION)

Scale: 0 5 10 20 30 miles

© RAND McNALLY & CO.   PRINTED IN U.S.A.

93-1

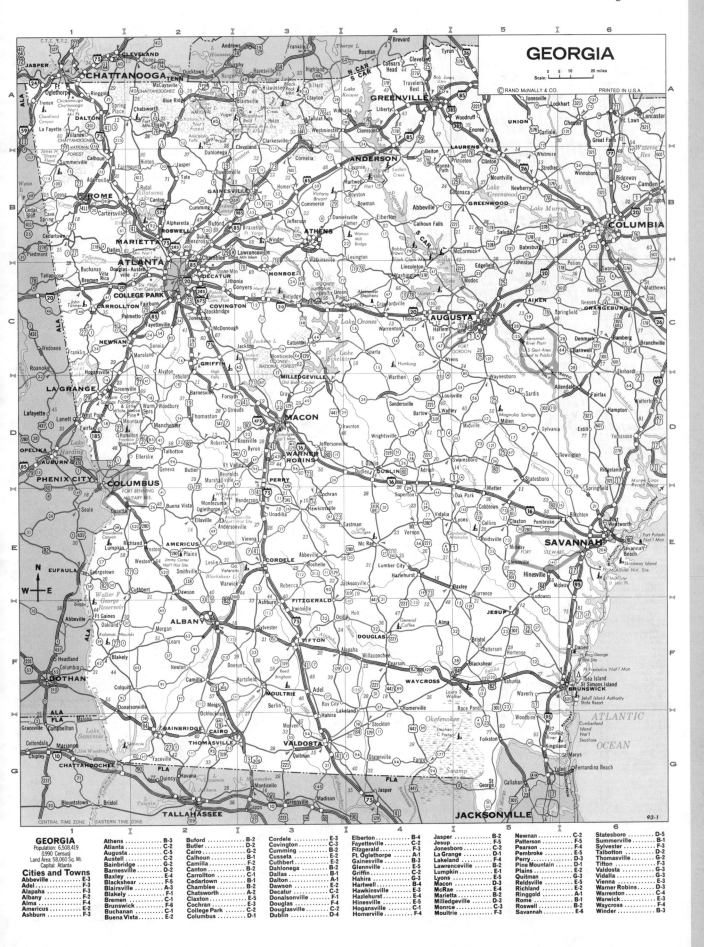

GEORGIA

Scale:

© RAND McNALLY & CO.   PRINTED IN U.S.A.

93-1

GEORGIA
Population: 6,508,419
(1990 Census)
Land Area: 58,060 Sq. Mi.
Capital: Atlanta

Cities and Towns

IDAHO

Population: 1,011,986
(1990 Census)
Land Area: 82,413 Sq. Mi.
Capital: Boise

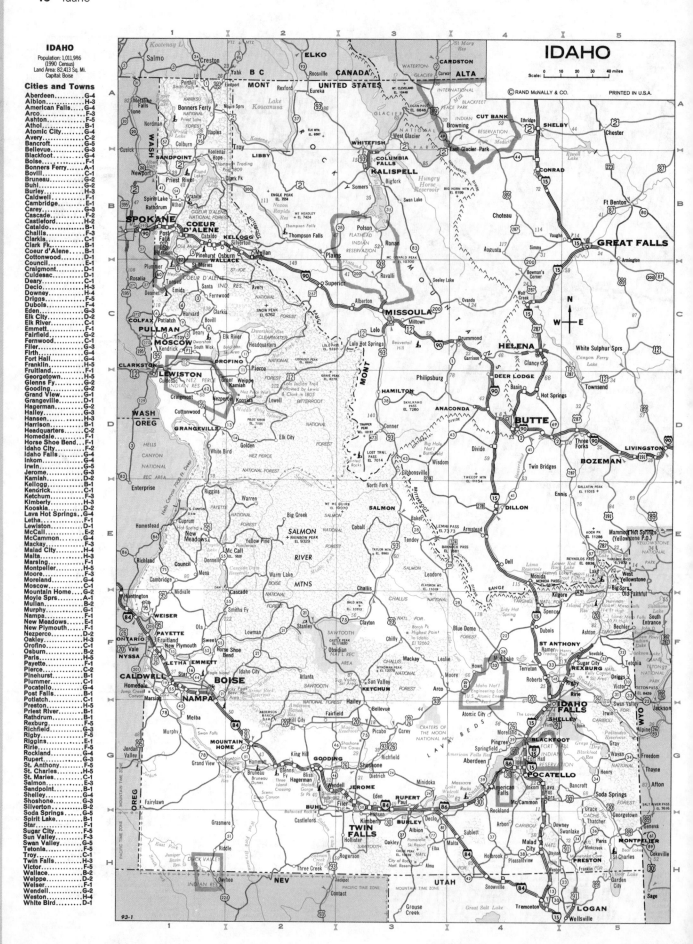

IDAHO

Scale 0 10 20 30 40 miles

© RAND McNALLY & CO.

PRINTED IN U.S.A.

93-1

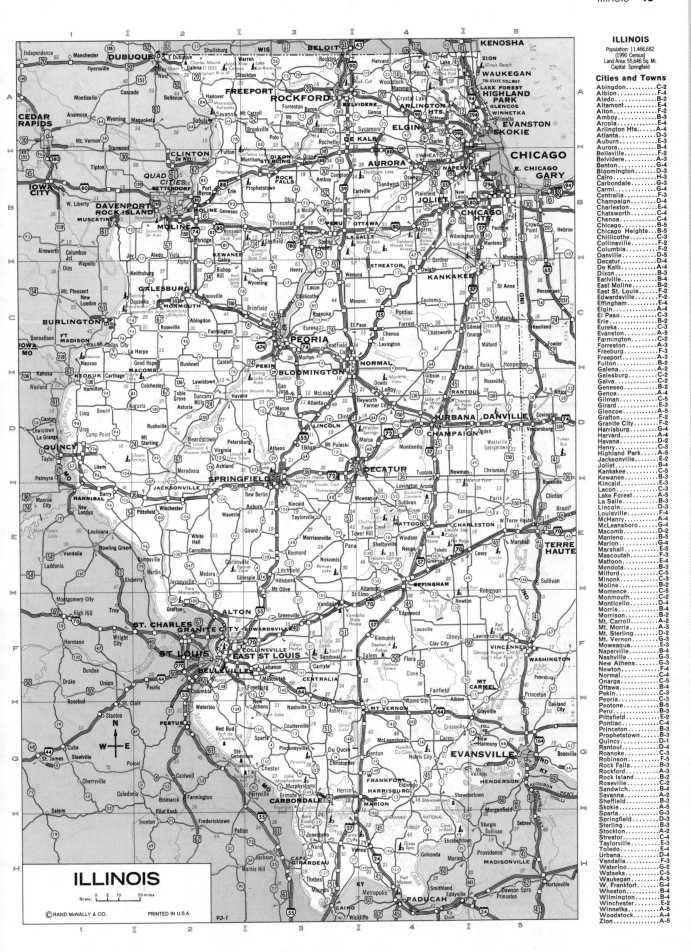

ILLINOIS

ILLINOIS
Population: 11,466,682
(1990 Census)
Land Area: 55,645 Sq. Mi.
Capital: Springfield

Cities and Towns

Abingdon..........C-2
Albion............F-4
Aledo.............B-2
Altamont..........E-4
Alton.............F-2
Amboy.............B-3
Arcola............E-4
Arlington Hts.....A-4
Atlanta...........D-3
Auburn............E-3
Aurora............B-4
Belleville........F-2
Belvidere.........A-3
Benton............G-4
Bloomington.......D-3
Cairo.............H-3
Carbondale........G-4
Carmi.............G-4
Centralia.........F-3
Champaign.........D-4
Charleston........E-4
Chatsworth........C-4
Chenoa............C-4
Chicago...........B-5
Chicago Heights...B-5
Chillicothe.......C-3
Collinsville......F-2
Columbia..........F-2
Danville..........D-5
Decatur...........D-4
De Kalb...........A-4
Dixon.............B-3
Earlville.........B-4
East Moline.......B-2
East St. Louis....F-2
Edwardsville......F-2
Effingham.........E-4
Elgin.............A-4
El Paso...........C-3
Erie..............B-2
Eureka............C-3
Evanston..........A-5
Farmington........C-2
Forreston.........A-3
Freeburg..........F-3
Freeport..........A-3
Fulton............B-2
Galena............A-2
Galesburg.........C-2
Galva.............C-2
Geneseo...........B-2
Genoa.............A-4
Gilman............C-5
Girard............E-3
Glencoe...........A-5
Grafton...........F-2
Granite City......F-2
Harrisburg........G-4
Harvard...........A-4
Havana............D-2
Henry.............C-3
Highland Park.....A-5
Jacksonville......E-2
Joliet............B-4
Kankakee..........B-5
Kewanee...........C-3
Kincaid...........E-3
Lacon.............C-3
Lake Forest.......A-5
La Salle..........B-3
Lincoln...........D-3
Louisville........F-4
McHenry...........A-4
McLeansboro.......G-4
Macomb............D-2
Manteno...........B-5
Marion............G-4
Marshall..........E-5
Mascoutah.........F-2
Mattoon...........E-4
Mendota...........B-3
Milford...........C-5
Minonk............C-3
Moline............B-2
Momence...........C-5
Monmouth..........C-2
Monticello........D-4
Morris............B-4
Morrison..........B-2
Mt. Carroll.......A-2
Mt. Morris........A-3
Mt. Sterling......D-2
Mt. Vernon........G-3
Moweaqua..........E-3
Naperville........B-4
Nashville.........G-3
New Athens........F-2
Newton............F-4
Normal............C-3
Onarga............C-5
Ottawa............B-4
Pekin.............C-3
Peoria............C-3
Peotone...........B-5
Peru..............B-3
Pittsfield........E-2
Pontiac...........C-4
Princeton.........B-3
Prophetstown......B-3
Quincy............D-1
Rantoul...........D-4
Roanoke...........C-3
Robinson..........F-5
Rock Falls........B-3
Rockford..........A-3
Rock Island.......B-2
Roseville.........C-2
Sandwich..........B-4
Savanna...........A-2
Sheffield.........B-3
Skokie............A-5
Sparta............G-3
Springfield.......D-3
Sterling..........B-3
Stockton..........A-2
Streator..........C-4
Taylorville.......E-3
Toledo............D-4
Urbana............D-4
Vandalia..........F-3
Waterloo..........G-2
Watseka...........C-5
Waukegan..........A-5
W. Frankfort......G-4
Wheaton...........B-4
Wilmington........B-4
Winchester........E-2
Winnetka..........A-4
Woodstock.........A-4
Zion..............A-5

Scale: 0 5 10   20 miles

©RAND McNALLY & CO.   PRINTED IN U.S.A.

93-1

INDIANA
Population: 5,564,228
(1990 Census)
Land Area: 35,936 Sq. Mi.
Capital: Indianapolis

Cities and Towns

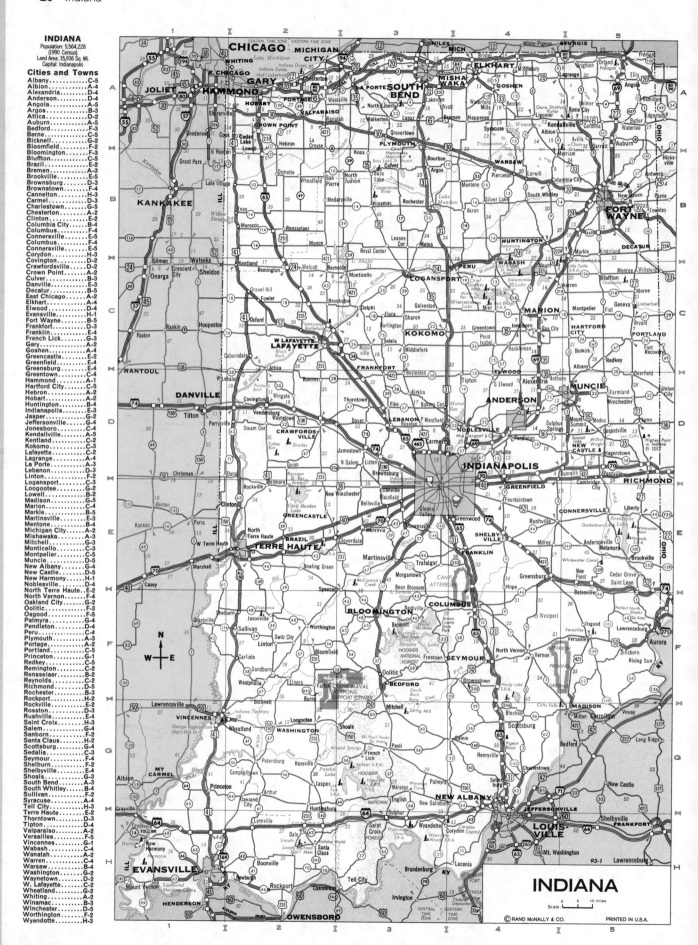

INDIANA

Scale: 0   5   10 miles

© RAND McNALLY & CO.   PRINTED IN U.S.A.

93-1

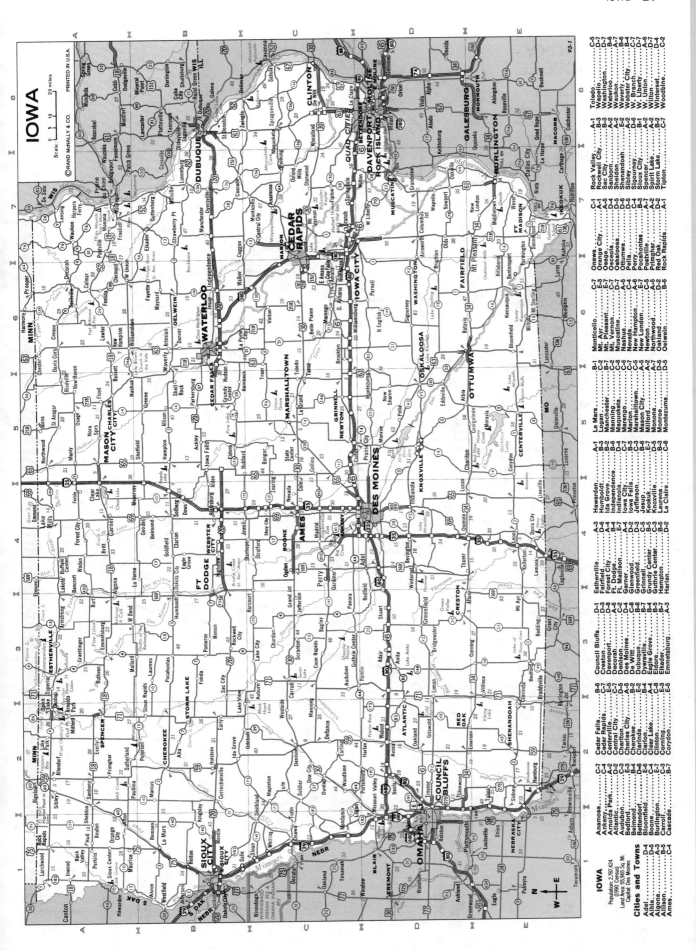

IOWA
Population, 2,787,424
1990 census
Land Area, 55,965 Sq. Mi.
Capital: Des Moines

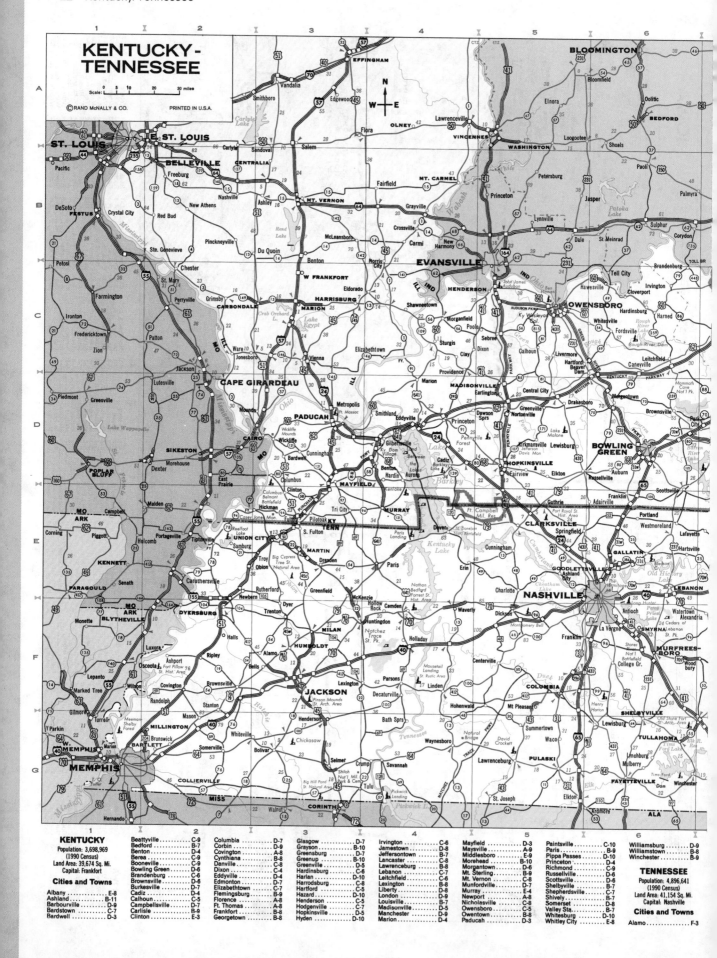

KENTUCKY

Population: 3,698,969
(1990 Census)
Land Area: 39,674 Sq. Mi.
Capital: Frankfort

Cities and Towns

Albany	E-8
Ashland	B-11
Barbourville	D-9
Bardstown	C-7
Bardwell	D-3

Beattyville	C-9
Bedford	B-7
Benton	D-4
Berea	C-9
Booneville	C-9
Bowling Green	D-6
Brandenburg	C-6
Brownsville	D-6
Burkesville	D-7
Cadiz	D-4
Calhoun	C-5
Campbellsville	C-7
Carlisle	B-9
Clinton	E-3

Columbia	D-7
Corbin	D-9
Covington	A-8
Cynthiana	B-8
Danville	C-8
Dixon	C-4
Eddyville	D-4
Edmonton	D-7
Elizabethtown	C-7
Flemingsburg	B-9
Florence	A-8
Ft. Thomas	A-8
Frankfort	B-8
Georgetown	B-8

Glasgow	D-7
Grayson	B-10
Greensburg	C-7
Greenup	B-10
Greenville	D-5
Hardinsburg	C-6
Harlan	D-10
Harrodsburg	C-8
Hartford	C-5
Hazard	D-10
Henderson	C-5
Hodgenville	C-7
Hopkinsville	D-4
Hyden	D-10

Irvington	C-6
Jamestown	D-8
Jeffersontown	B-7
Lancaster	C-8
Lawrenceburg	B-8
Lebanon	C-7
Leitchfield	C-6
Lexington	B-8
Liberty	C-8
London	D-9
Louisville	B-7
Madisonville	C-4
Manchester	D-9
Marion	D-4

Mayfield	D-3
Maysville	A-9
Middlesboro	E-9
Morehead	B-10
Morganfield	C-4
Mt. Sterling	B-9
Mt. Vernon	C-9
Munfordville	D-7
Murray	E-4
Newport	A-8
Nicholasville	C-8
Owensboro	C-5
Owentown	B-8
Paducah	D-3

Paintsville	C-10
Paris	B-9
Pippa Passes	D-10
Princeton	D-4
Richmond	C-9
Russellville	D-6
Scottsville	D-6
Shelbyville	B-7
Shepherdsville	C-7
Shively	B-7
Somerset	D-8
Valley Sta.	B-7
Whitesburg	D-10
Whitley City	E-8

Williamsburg	D-9
Williamstown	B-8
Winchester	B-9

TENNESSEE

Population: 4,896,641
(1990 Census)
Land Area: 41,154 Sq. Mi.
Capital: Nashville

Cities and Towns

Alamo	F-3

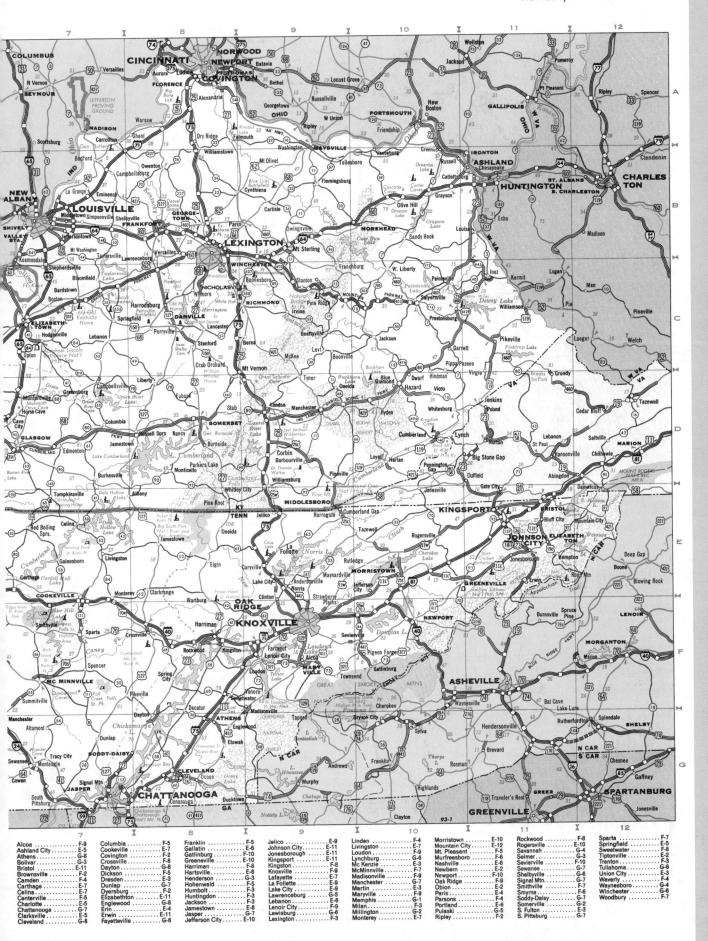

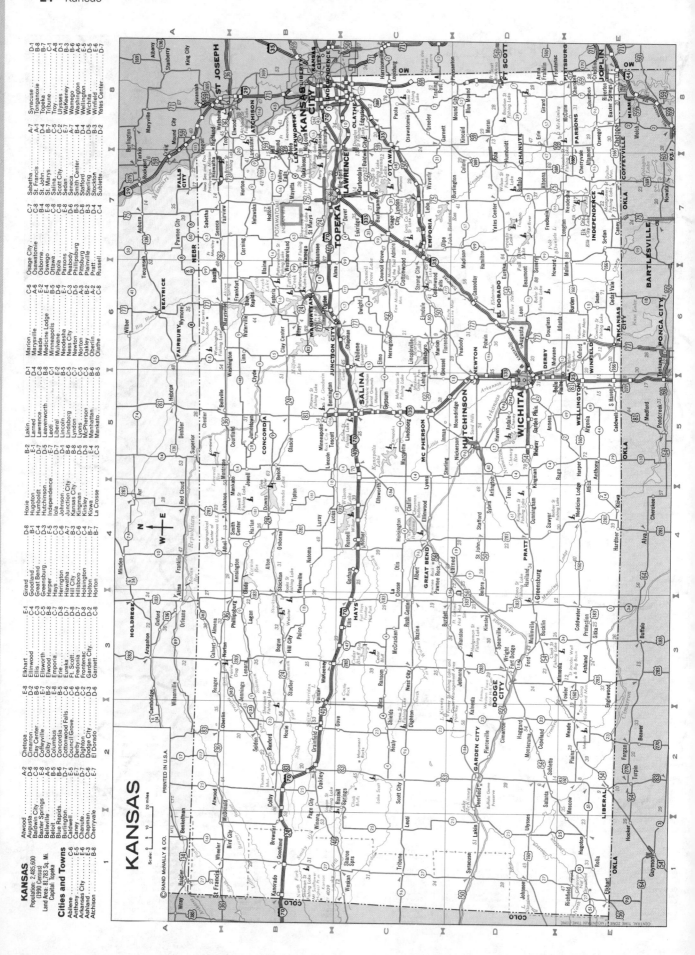

KANSAS

Scale: 5 10 20
30 miles

©RAND McNALLY & CO.
PRINTED IN U.S.A.

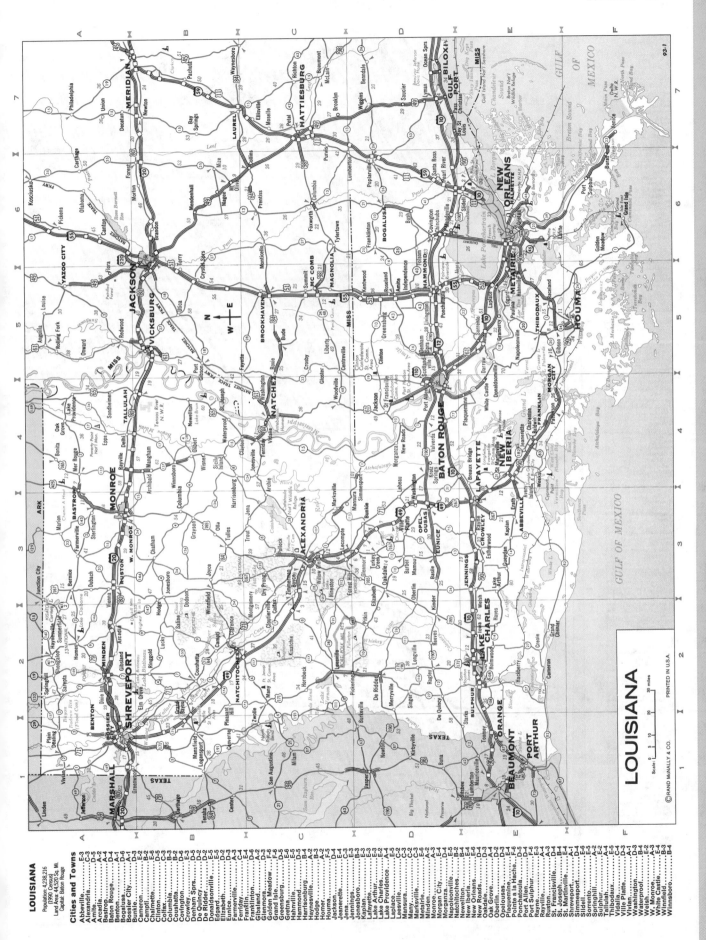

93-1

LOUISIANA

Scale: miles

0 5 10 20 30

© RAND McNALLY & CO.   PRINTED IN U.S.A.

LOUISIANA

Population: 4,238,216
(1990 Census)
Land Area: 44,520 Sq. Mi.
Capital: Baton Rouge

Cities and Towns

Abbeville.........E-3
Alexandria........C-3
Amite.............D-5
Arcadia...........B-2
Bastrop...........A-4
Baton Rouge.......D-4
Benton............A-1
Bogalusa..........D-5
Bossier City......A-1
Bunkie............C-3
Cameron...........E-2
Chalmette.........D-5
Clinton...........D-4
Colfax............C-3
Columbia..........B-3
Coushatta.........B-2
Covington.........D-5
Crowley...........E-3
De Quincy.........D-2
De Ridder.........D-2
Denham Sprs.......D-5
Donaldsonville....D-4
Edgard............D-5
Elizabeth.........D-3
Eunice............D-3
Farmerville.......A-3
Ferriday..........C-4
Franklin..........E-4
Franklinton.......D-5
Gibsland..........B-2
Glenmora..........D-3
Golden Meadow.....F-6
Grand Isle........F-6
Greensburg........D-5
Gretna............D-5
Hahnville.........D-5
Hammond...........D-5
Harrisonburg......C-4
Haynesville.......A-2
Hodge.............B-3
Homer.............A-2
Houma.............F-5
Jackson...........D-4
Jeanerette........E-4
Jena..............C-3
Jennings..........E-3
Jonesboro.........B-3
Kaplan............E-3
Lafayette.........E-3
Lake Arthur.......E-3
Lake Charles......E-2
Lake Providence...A-4
Laplace...........D-5
Leesville.........C-2
Mansfield.........B-1
Mansura...........C-3
Many..............C-2
Marksville........C-3
Metairie..........D-5
Minden............A-2
Monroe............B-3
Morgan City.......E-4
Napoleonville.....E-4
Natchitoches......C-2
Newellton.........B-4
New Iberia........E-3
New Orleans.......D-5
New Roads.........D-4
Oakdale...........D-3
Oak Grove.........A-4
Oberlin...........D-3
Opelousas.........D-3
Plaquemine........D-4
Pointe a la Hache.F-6
Ponchatoula.......D-5
Port Allen........D-4
Port Sulphur......F-6
Rayne.............E-3
Rayville..........B-4
Ruston............B-2
St. Francisville..D-4
St. Joseph........B-4
St. Martinville...E-3
Shreveport........A-1
Simmesport........C-3
Slidell...........D-5
Sorrento..........D-4
Springhill........A-1
Sulphur...........E-2
Tallulah..........B-4
Thibodaux.........E-4
Vidalia...........C-4
Ville Platte......D-3
Washington........D-3
Waterproof........B-4
Welsh.............E-3
W. Monroe.........B-3
Winnfield.........B-3
Winnsboro.........B-4

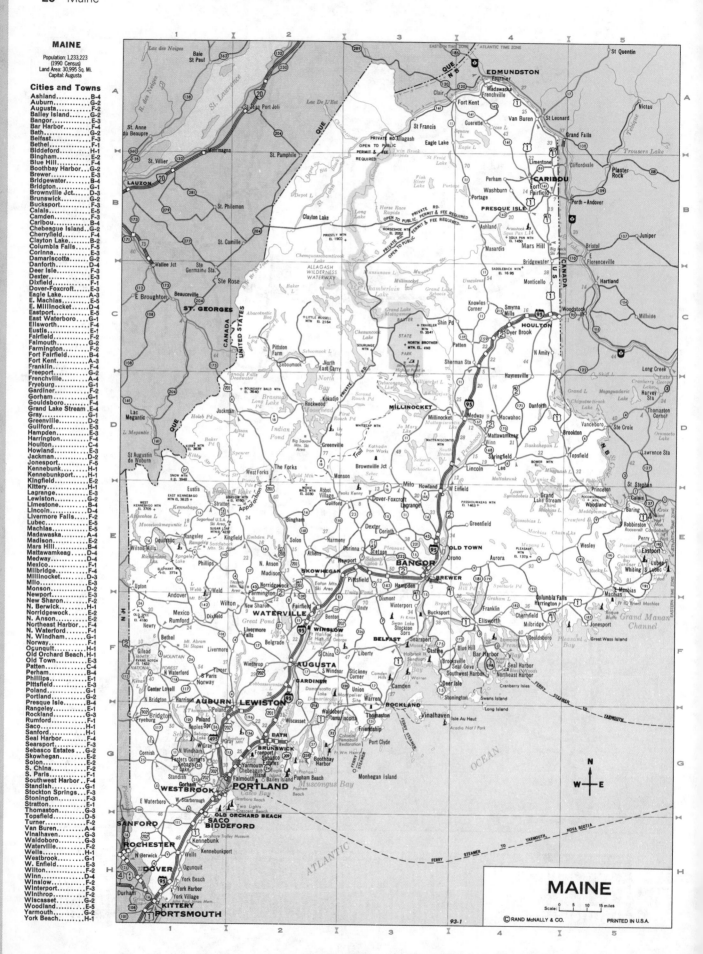

MAINE

Scale: 0 5 10 15 miles

© RAND McNALLY & CO.

PRINTED IN U.S.A.

93-1

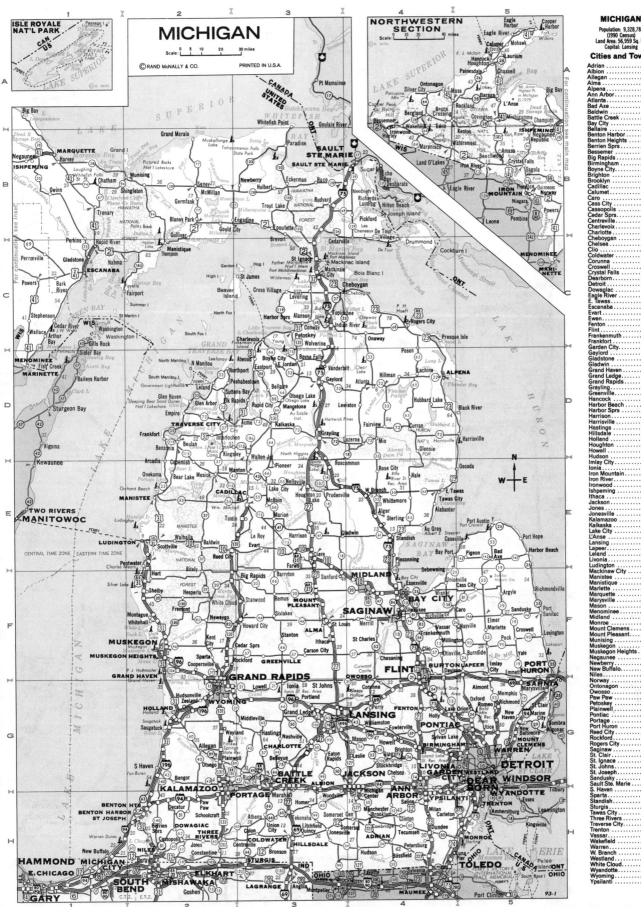

MICHIGAN

Population: 9,328,784
(1990 Census)
Land Area: 56,959 Sq. Mi.
Capital: Lansing

Cities and Towns

Adrian	H-4
Albion	H-3
Allegan	G-2
Alma	F-3
Alpena	D-4
Ann Arbor	G-4
Atlanta	D-4
Bad Axe	E-5
Baldwin	E-2
Battle Creek	G-3
Bay City	F-4
Bellaire	D-3
Benton Harbor	H-2
Benton Heights	H-2
Berrien Sprs	H-2
Bessemer	A-4
Big Rapids	F-3
Birmingham	G-5
Boyne City	D-3
Brighton	G-4
Brooklyn	H-4
Cadillac	E-3
Calumet	A-5
Caro	F-4
Cass City	F-5
Cassopolis	H-2
Cedar Sprs	F-2
Centreville	H-3
Charlevoix	C-3
Charlotte	G-3
Cheboygan	C-4
Chelsea	G-4
Clio	F-4
Coldwater	H-3
Corunna	G-4
Croswell	F-5
Crystal Falls	B-5
Dearborn	G-5
Detroit	H-5
Dowagiac	H-2
Eagle River	A-5
E. Tawas	E-4
Escanaba	C-1
Evart	E-3
Ewen	A-4
Fenton	G-4
Flint	F-4
Frankenmuth	F-4
Frankfort	D-2
Garden City	G-5
Gaylord	D-3
Gladstone	C-1
Gladwin	E-3
Grand Haven	F-2
Grand Ledge	G-3
Grand Rapids	G-2
Grayling	E-3
Greenville	F-3
Hancock	A-5
Harbor Beach	E-5
Harbor Sprs	C-3
Harrison	E-3
Harrisville	D-5
Hastings	G-3
Hillsdale	H-3
Holland	G-2
Houghton	A-5
Howell	G-4
Hudson	H-4
Imlay City	G-5
Ionia	G-3
Iron Mountain	B-5
Iron River	A-5
Ironwood	B-4
Ishpeming	B-1
Ithaca	F-3
Jackson	G-4
Jones	H-2
Jonesville	H-3
Kalamazoo	G-2
Kalkaska	D-3
Lake City	E-3
L'Anse	A-5
Lansing	G-3
Lapeer	F-5
Leland	D-2
Livonia	G-5
Ludington	E-2
Mackinaw City	C-3
Manistee	E-2
Manistique	C-2
Marlette	F-5
Marquette	B-1
Marysville	G-5
Mason	G-3
Menominee	D-1
Midland	F-4
Monroe	H-5
Mount Clemens	G-5
Mount Pleasant	F-3
Munising	B-2
Muskegon	F-2
Muskegon Heights	F-2
Negaunee	B-1
Newberry	B-3
New Buffalo	H-1
Niles	H-2
Norway	B-5
Ontonagon	A-4
Owosso	G-4
Paw Paw	H-2
Petoskey	C-3
Plainwell	G-2
Pontiac	G-5
Portage	G-2
Port Huron	G-5
Reed City	E-3
Rockford	F-2
Rogers City	C-4
Saginaw	F-4
St. Clair	G-5
St. Ignace	C-3
St. Johns	G-3
St. Joseph	H-2
Sandusky	F-5
Sault Ste. Marie	B-4
S. Haven	H-2
Sparta	F-2
Standish	E-4
Sturgis	H-3
Tawas City	E-4
Three Rivers	H-3
Traverse City	D-2
Trenton	H-5
Vassar	F-4
Wakefield	A-4
Warren	G-5
W. Branch	E-4
Westland	G-5
White Cloud	F-2
Wyandotte	H-5
Wyoming	G-2
Ypsilanti	H-4

MINNESOTA

Population: 4,387,029
(1990 Census)
Land Area: 79,548 Sq. Mi.
Capital: St. Paul

Cities and Towns

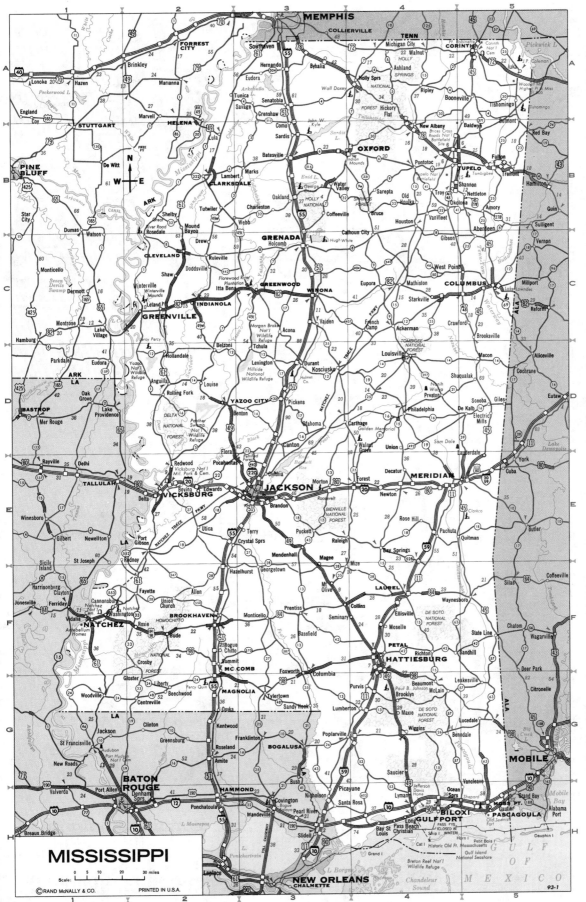

93-1

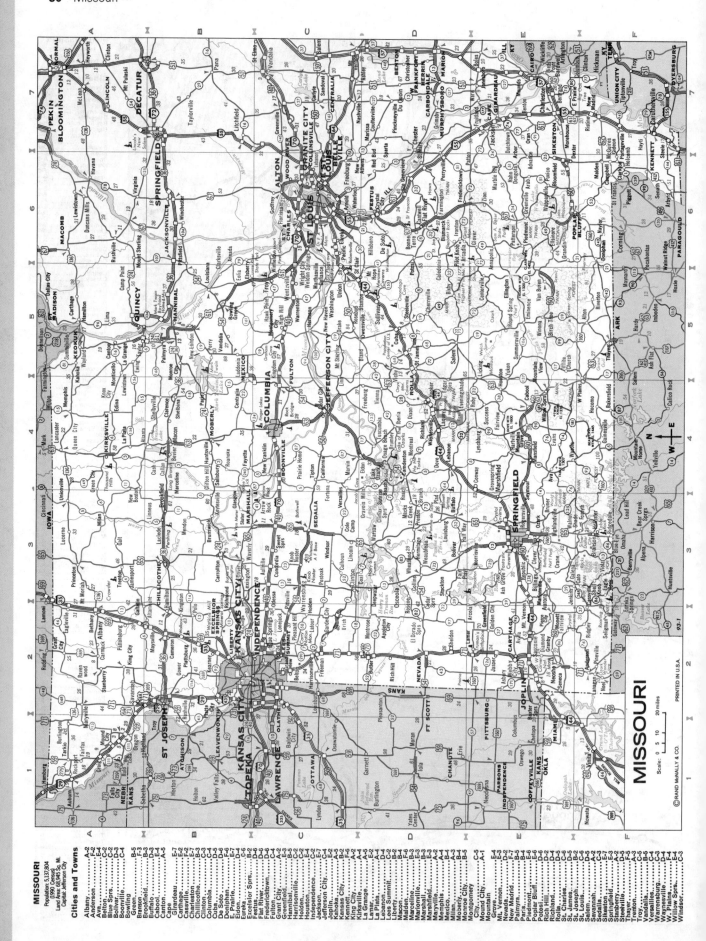

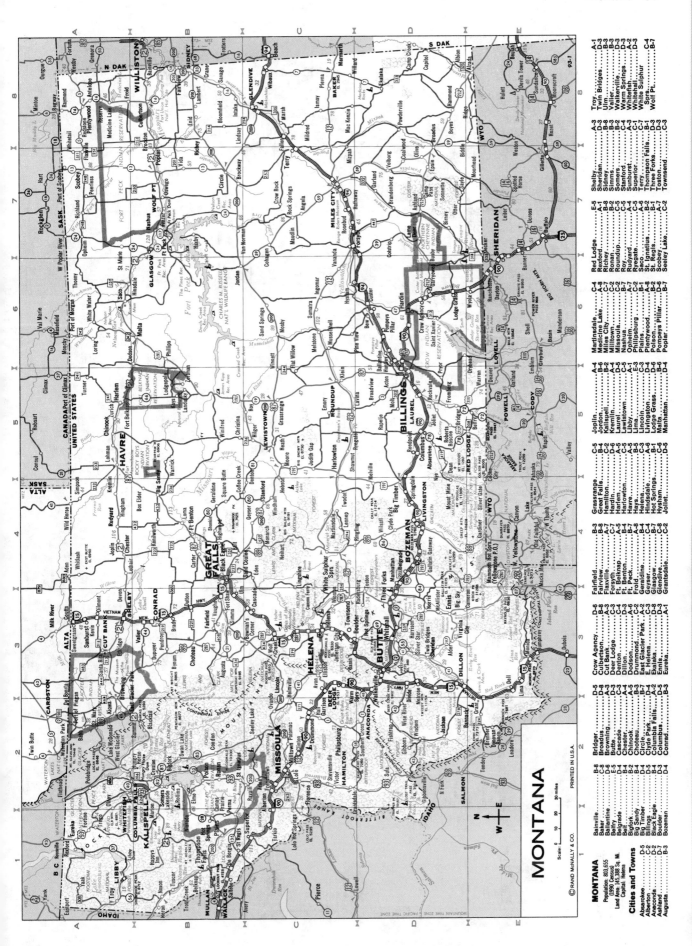

MONTANA

MONTANA
Population: 803,655
(1990 Census)
Land Area: 145,388 Sq. Mi.
Capital: Helena

Scale:
0 10 20 30 miles

© RAND McNALLY & CO.   PRINTED IN U.S.A.

MOUNTAIN TIME ZONE & PACIFIC TIME ZONE

N
W—E
S

93-1

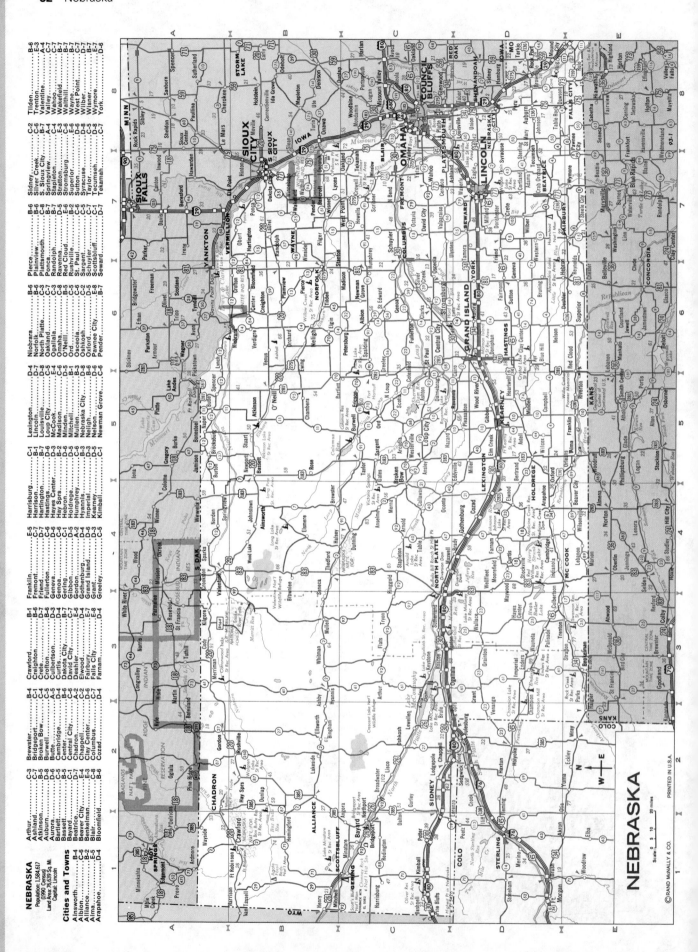

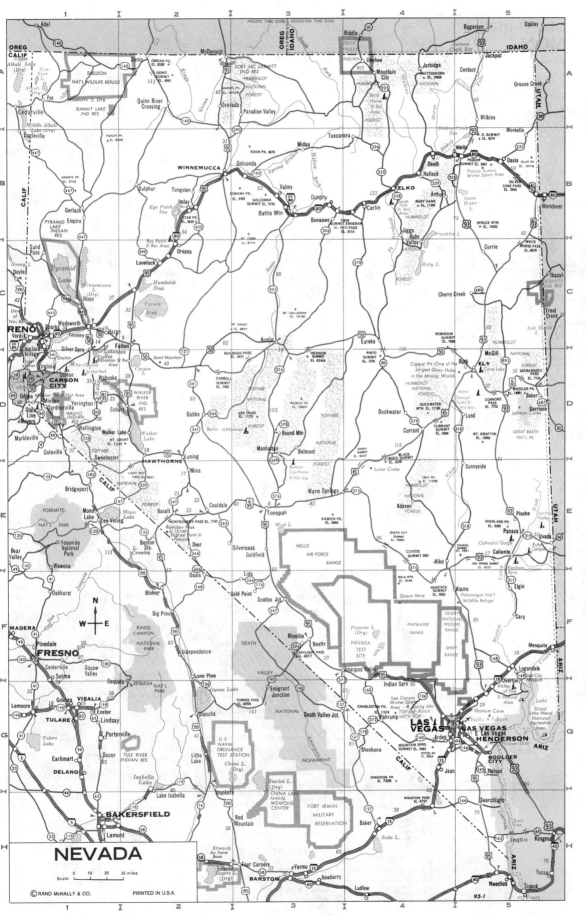

NEVADA

NEVADA
Population: 1,206,152
(1990 Census)
Land Area: 109,895 Sq. Mi.
Capital: Carson City

NEW HAMPSHIRE-VERMONT

Scale: 0   5   10 miles

© RAND McNALLY & CO.   PRINTED IN U.S.A.

N. HAMPSHIRE

Population: 1,113,915
(1990 Census)
Land Area: 8,992 Sq. Mi.
Capital: Concord

Cities and Towns

Ashland	E-4
Berlin	C-5
Bristol	E-4
Canaan	E-4
Center Ossipee	E-5
Charlestown	F-3
Chocorua	D-5
Claremont	F-3
Colebrook	B-5
Concord	F-5
Conway	D-5
Derry	G-5
Dover	F-6
Epping	F-5
Franklin	E-4
Glen	D-5
Gorham	C-5
Groveton	B-5
Hampton	G-6
Hampton Beach	G-6
Hanover	E-3
Haverhill	D-4
Henniker	F-4
Hillsborough	F-4
Hudson	G-5
Jaffrey	G-4
Keene	G-3
Laconia	E-5
Lancaster	C-5
Lebanon	E-4
Lincoln	D-4
Littleton	C-4
Manchester	G-5
Meredith	E-5
Milford	G-5
Nashua	G-5
Newport	F-3
North Conway	D-5
N. Stratford	B-4
N. Woodstock	D-4
Peterborough	G-4
Pittsfield	F-5
Plymouth	E-4
Portsmouth	F-6
Rochester	F-6
Rye Beach	G-6
Salem	G-5
Troy	G-3
Whitefield	C-4
Winchester	G-3
Woodsville	D-4

VERMONT

Population: 564,964
(1990 Census)
Land Area: 9,273 Sq. Mi.
Capital: Montpelier

Cities and Towns

Arlington	F-1
Barre	C-3
Barton	B-3
Bellows Falls	F-3
Bennington	G-1
Bethel	D-2
Brandon	D-2
Brattleboro	G-3
Bristol	D-2
Burlington	C-1
Cambridge	B-2
Chester	F-3
Enosburg Falls	B-2
Essex Jct.	C-2
Fair Haven	E-1
Grafton	F-3
Hardwick	C-3
Highgate Sprs.	A-2
Island Pond	B-4
Ludlow	F-2
Lyndonville	C-4
Middlebury	D-1
Montpelier	C-3
Morrisville	C-3
Newport	B-3
Norwich	E-3
Poultney	E-1
Randolph	D-2
Richford	A-2
Rutland	E-2
St. Albans	B-2
St. Johnsbury	C-4
Shelburne	C-1
S Burlington	C-1
Springfield	F-3
Swanton	B-2
Vergennes	D-1
Waterbury	C-2
Wells River	C-3
White River Jct.	E-3
Windsor	E-3
Winooski	C-1
Woodstock	E-3

93-1

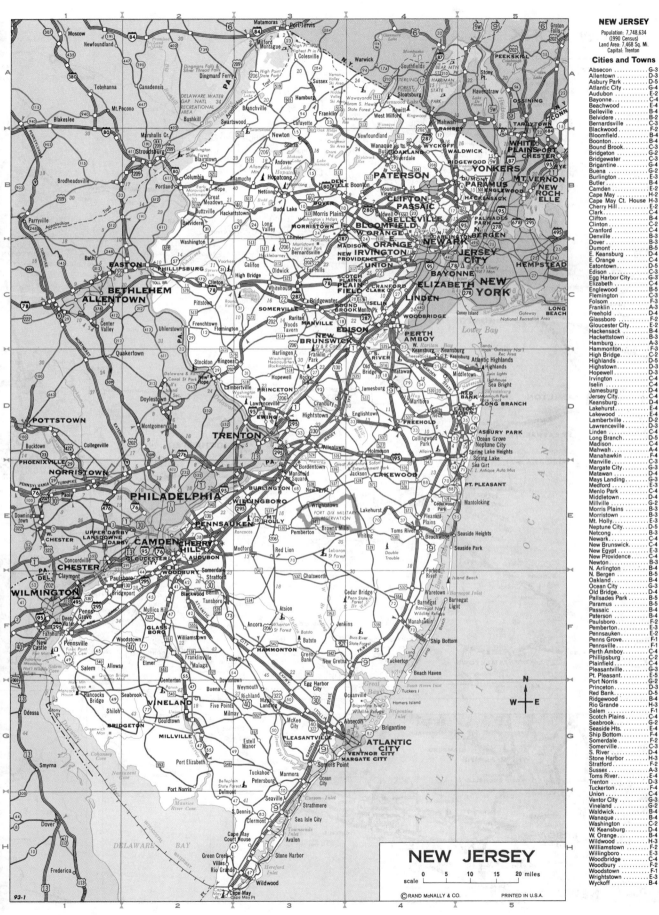

NEW JERSEY

Population: 7,748,634
(1990 Census)
Land Area: 7,468 Sq. Mi.
Capital: Trenton

Cities and Towns

Absecon G-3
Allentown D-3
Asbury Park D-5
Atlantic City G-4
Audubon E-2
Bayonne C-4
Beachwood E-4
Belleville B-4
Belvidere B-2
Bernardsville C-3
Blackwood F-2
Bloomfield B-4
Boonton B-4
Bound Brook C-3
Bridgeton G-2
Bridgewater C-3
Brigantine G-4
Buena G-2
Burlington E-3
Butler B-4
Camden E-2
Cape May H-2
Cape May Ct. House . . . H-3
Cherry Hill E-2
Clark C-4
Clifton B-4
Clinton C-2
Cranford C-4
Denville B-3
Dover B-3
Dumont B-5
E. Keansburg D-4
E. Orange C-4
Eatontown D-5
Edison C-3
Egg Harbor City G-3
Elizabeth C-4
Englewood B-5
Flemington C-3
Folsom F-3
Franklin A-3
Freehold D-4
Glassboro F-2
Gloucester City E-2
Hackensack B-4
Hackettstown B-3
Hamburg A-3
Hammonton F-3
High Bridge C-2
Highlands D-5
Hightstown D-3
Hopewell D-3
Irvington C-4
Iselin C-4
Jamesburg D-4
Jersey City C-4
Keansburg D-4
Lakehurst E-4
Lakewood E-4
Lambertville D-3
Lawrenceville D-3
Linden C-4
Long Branch D-5
Madison C-4
Mahwah A-4
Manahawkin F-4
Manville C-3
Margate City G-3
Matawan D-4
Mays Landing G-3
Medford E-3
Menlo Park C-4
Middletown D-4
Millville G-2
Morris Plains B-3
Morristown B-3
Mt. Holly E-3
Neptune City D-5
Netcong B-3
Newark C-4
New Brunswick C-4
New Egypt E-4
New Providence C-4
Newton B-3
N. Arlington C-4
N. Bergen B-5
Oakland B-4
Ocean City G-3
Old Bridge D-4
Palisades Park B-5
Paramus B-5
Passaic B-4
Paterson B-4
Paulsboro F-2
Pemberton E-3
Pennsauken E-2
Penns Grove F-1
Pennsville F-1
Perth Amboy C-4
Phillipsburg C-2
Plainfield C-4
Pleasantville G-3
Pt. Pleasant E-5
Port Norris G-2
Princeton D-3
Red Bank D-5
Ridgewood B-4
Rio Grande H-3
Salem F-1
Scotch Plains C-4
Seabrook G-2
Seaside Hts. E-4
Ship Bottom F-4
Somerdale E-2
Somerville C-3
S. River D-4
Stone Harbor H-3
Stratford E-2
Sussex A-3
Toms River E-4
Trenton D-3
Tuckerton F-4
Union C-4
Ventor City G-3
Vineland G-2
Waldwick B-4
Wanaque B-4
Washington B-2
W. Keansburg D-4
W. Orange C-4
Wildwood H-3
Williamstown F-2
Willingboro E-3
Woodbridge C-4
Woodbury F-2
Woodstown F-1
Wrightstown E-3
Wyckoff B-4

NEW YORK

Scale: 0 5 10   20 miles

© RAND McNALLY & CO.   PRINTED IN U.S.A.

Nottawasagu Bay
Collingwood
Coldwater
ORILLIA
Lake Simcoe
BARRIE
Beaverton
Sutton
LINDSAY
Manchester
PETERBOROUGH
Rice L.
Havelock
Marmora
Madoc
Kaladar
Sharbot Lake
Perth
Smiths Falls
Rideau L.

Primrose
Schomberg
Orangeville
Caledon
Woodbridge
BRAMPTON
GUELPH
Morriston
Burlington
DUNDAS
Peters Cor's
HAMILTON
ST CATHARINES
Allanburg
Bismarck
NIAGARA FALLS
WELLAND
Jarvis
Dunnville
Long Point Bay

TORONTO
OSHAWA
New Castle
Port Hope
Bloomfield
TRENTON
BELLEVILLE
Napanee
KINGSTON
Wolfe Island
Cape Vincent
Clayton
Chaumont
WATERTOWN
Sackets Hbr.
Henderson
Adams
Southwick Beach
Pulaski
Mexico Bay
Selkirk Shores
New Haven
OSWEGO
Mexico
Maple View
Hannibal
Red Creek
FULTON
Baldwinsville
Cicero

LAKE ONTARIO

CANADA UNITED STATES
Youngstown
Fort Niagara
Lewiston
LOCKPORT
NIAGARA FALLS
Niagara Falls
BUFFALO
LACKAWANNA
CHEEKTOWAGA
DEPEW
Orchard Park
East Aurora
Hamburg
Evans Cen
Evangola
Silver Cr
DUNKIRK
Fredonia
Brocton
Westfield
Mayville
Ripley
ERIE
Chautauqua
Panama
JAMESTOWN
Frewsburg
Clymer
Falconer
Cutting

Golden Hill
Olcott
Ridgeway
Medina
Albion
Carlton
Childs
Hilton
Clarkson
Brockport
Webster
Williamson
Alton
Sodus Point
Fair Haven Beach
Wolcott
N. Victor
Fort Brewerton

ROCHESTER
Henrietta
NEWARK
Lyons
Clyde
SYRACUSE
AUBURN
La Fayette
Skaneateles
Seneca Falls
GENEVA
Aurora
Moravia
Genoa
Homer
CORTLAND
McGraw
Labrador Ski Area

BATAVIA
Le Roy
Pavilion Cen
Avon
Victor
Manchester
CANANDAIGUA
Waterloo
Ovid
Locke

Alabama
Tonawanda Indian Res
Darien
Attica
Pavilion
Warsaw
Perry
Varysburg
Chafee
Arcade
Springville
Gowanda
Dayton
Cadiz
Franklinville
Belfast
Little Valley
Great Valley
SALAMANCA
ALLEGHENY INDIAN RES.
OLEAN
Bolivar
Portville
Ceres

E Avon
Lakeville
Livonia
Geneseo
Nunda
Dansville
Wayland
Naples
Penn Yan
Cohocton
Garwoods
Caneadea
Belvidere
Belmont
Andover
Wellsville
Cuba
Greenwood
Jasper
Addison

HORNELL
Canisteo
Bath
Savona
Painted Post
CORNING
ELMIRA
Horseheads
Van Etten
Spencer
Candor
BINGHAMTON
ENDICOTT
Owego
Waverly

Watkins Glen
Montour Falls
ITHACA
Dryden
Harford
Richford
Whitney Pt
Towanda
Wyalusing

PA

N
W — E
S

Spring Valley
Suffern
PATERSON
CLIFTON
E ORANGE
IRVINGTON
NEWARK
BAYONNE
ELIZABETH
JERSEY CITY
YONKERS
MT VERNON
NEW ROCHELLE
WHITE PLAINS
STAMFORD
NORWALK
LONG ISLAND SOUND
Greenport
Orient Beach
Southold
Shelter Island
Mattituck
Cutchogue
Sag Harbor
Montauk Point
Montauk
Amagansett
E Hampton
Southampton
Hampton Bays
Quogue
Westhampton Beach
Patchogue
Sayville
Riverhead
Port Jefferson
Smithtown
Coram
Medford Sta
Central Islip
Bay Shore
Fire Island Nat'l Seashore
Great South Bay
Robert Moses
Long Beach
Jones Beach
OCEANSIDE
FREEPORT
MERRICK
Hicksville
JERICHO
HUNTINGTON
Oyster Bay
GLEN COVE
Great Neck
Long Island City
Port Washington
Block Island Sound

NEW YORK

ATLANTIC OCEAN

© R MN. & CO.

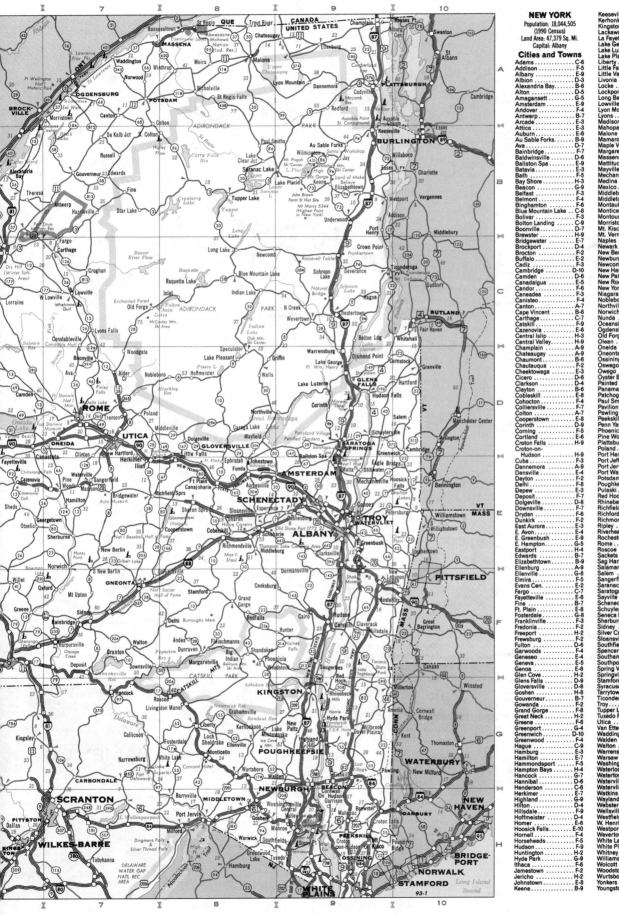

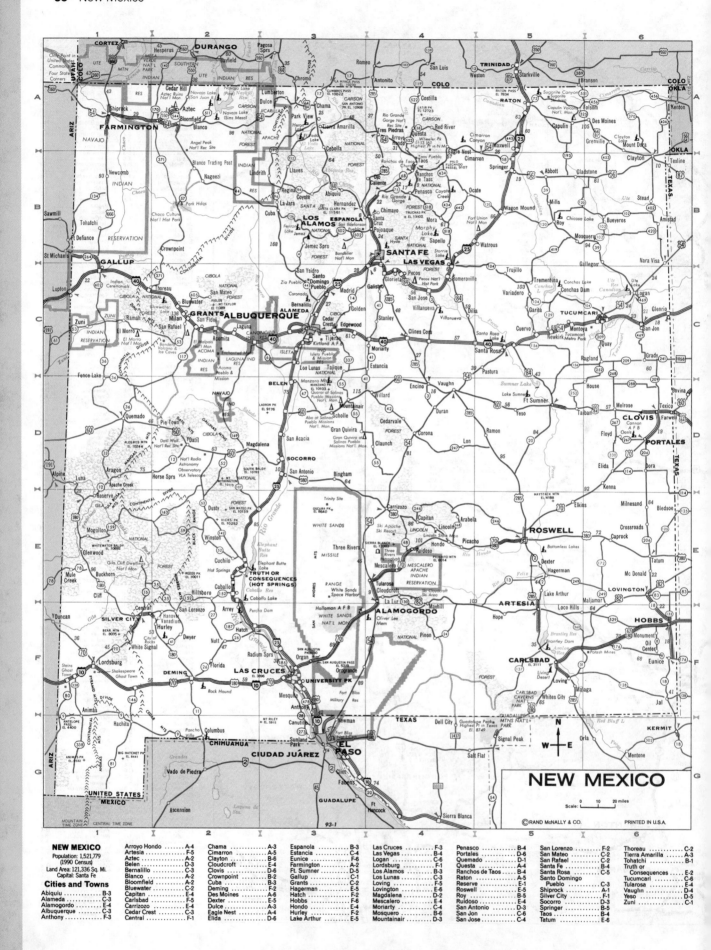

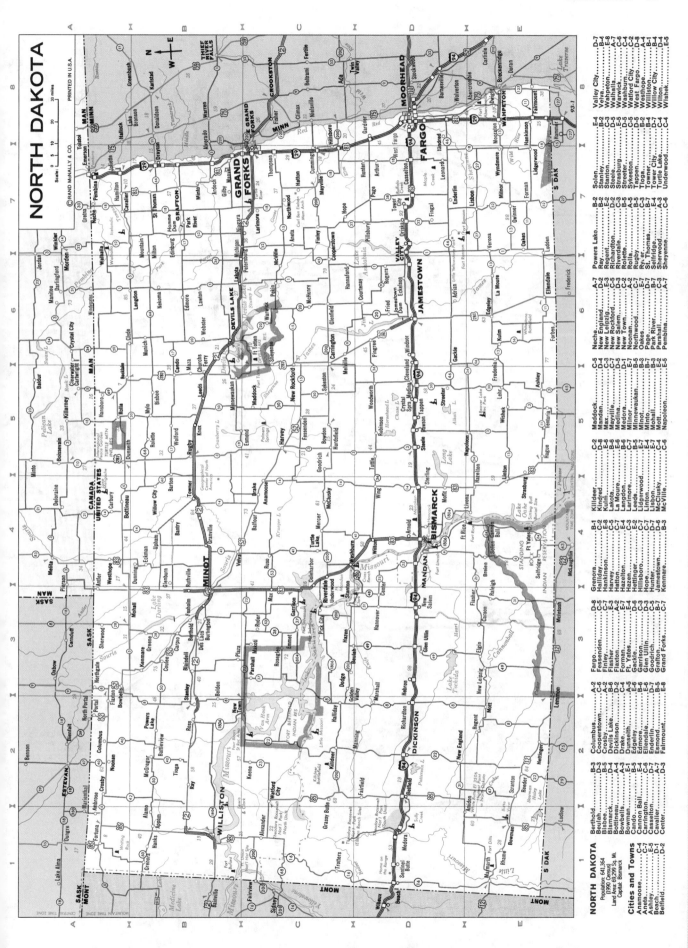

NORTH DAKOTA

© RAND McNALLY & CO.

PRINTED IN U.S.A.

Scale:
0 5 10 20 30 miles

NORTH DAKOTA

Population: 641,364
(1990 Census)
Land Area: 69,299 Sq. Mi.
Capital: Bismarck

Cities and Towns

Anamoose	C-4
Aneta	C-7
Ashley	E-6
Beach	D-1
Belfield	D-2
Berthold	B-3
Beulah	D-3
Bisbee	B-5
Bismarck	D-4
Bowbells	A-3
Bowman	E-1
Cando	B-5
Cannon Ball	D-4
Carrington	C-6
Cavalier	A-7
Center	D-3
Columbus	B-3
Cooperstown	D-3
Crosby	B-5
Devils Lake	A-3
Dickinson	D-5
Drake	D-2
Dunseith	B-5
Edgeley	E-4
Edmore	B-6
Ellendale	E-8
Enderlin	C-6
Fairmount	D-3
Fargo	A-2
Fessenden	D-3
Finley	B-2
Flasher	A-2
Flaxton	D-2
Ft. Yates	E-6
Gackle	E-6
Garrison	B-6
Glen Ullin	B-6
Goodrich	D-3
Grand Forks	C-7
Grenora	D-8
Halliday	C-7
Hankinson	B-5
Harvey	E-3
Hatton	E-4
Hazen	C-6
Hettinger	E-4
Hillsboro	D-6
Hope	D-3
Hunter	C-5
Jamestown	D-7
Kenmare	C-7
Killdeer	B-1
Kindred	C-2
Kulm	C-5
La Moure	C-7
Langdon	C-3
Larimore	E-4
Leeds	E-2
Lidgerwood	C-7
Lisbon	B-7
Linton	D-7
Maddock	C-2
Mandan	B-6
Max	E-6
Mayville	E-6
Medina	B-6
Medora	E-4
Michigan	B-5
Minnewaukan	E-7
Minot	E-4
Minto	E-4
Mott	C-4
Napoleon	C-6
Neche	C-5
New England	A-7
New Leipzig	E-3
New Rockford	D-3
New Salem	D-5
New Town	C-2
Northwood	B-5
Oakes	E-7
Page	D-7
Park River	E-7
Parshall	C-4
Pembina	E-5
Powers Lake	B-2
Ray	B-2
Regent	B-2
Richardton	D-2
Riverdale	C-3
Rolette	B-5
Rolla	B-2
Rugby	C-3
Ryder	D-7
St. Thomas	B-7
Selfridge	E-4
Sherwood	A-3
Sheyenne	C-6
Solen	B-2
Stanley	B-2
Stanton	B-2
Strasburg	D-2
Streeter	C-3
Sykeston	B-5
Tioga	B-2
Tower City	C-3
Towner	C-3
Turtle Lake	E-4
Underwood	A-3
Valley City	C-6
Wahpeton	E-4
Walhalla	C-3
Warwick	D-5
Washburn	E-5
Watford City	C-4
West Fargo	C-2
Westhope	B-2
Williston	A-4
Willow City	B-1
Wilton	D-7
Wishek	C-4
	D-5

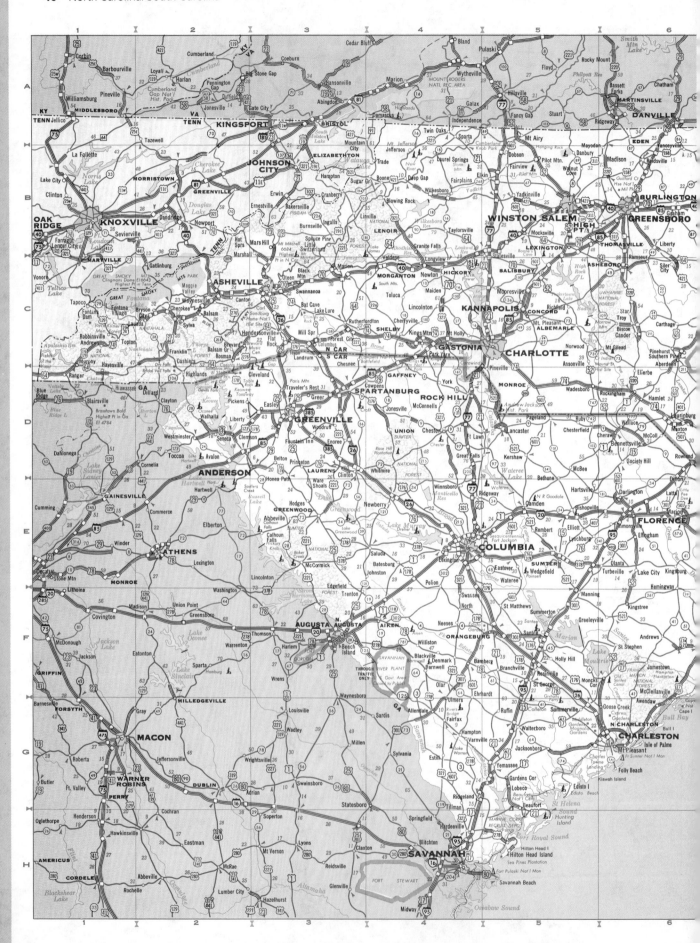

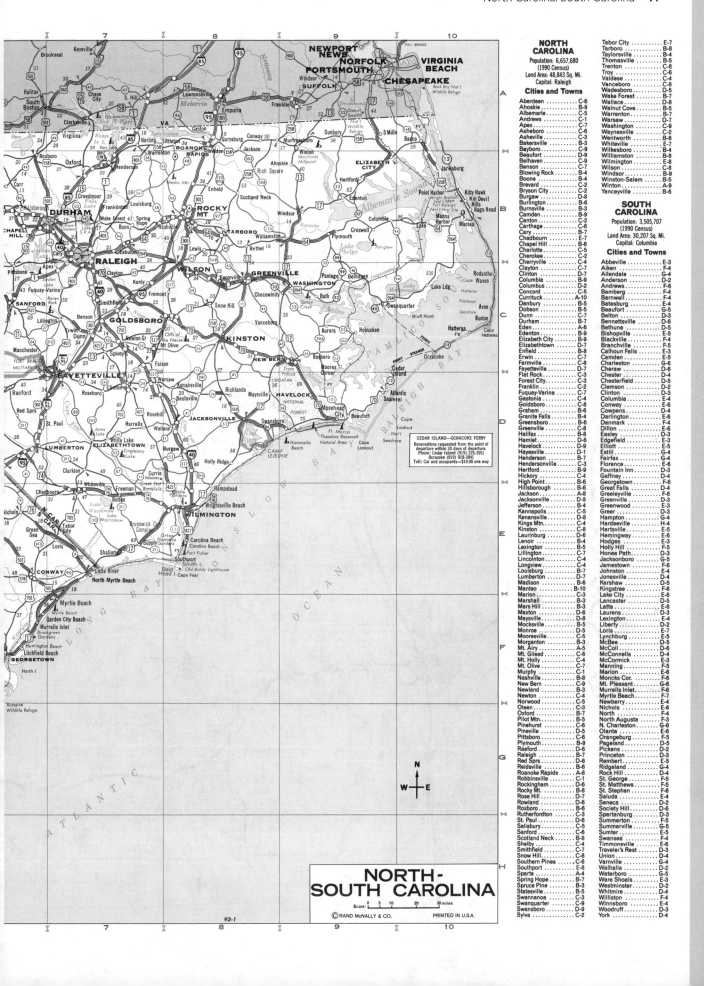

NORTH CAROLINA

Population: 6,657,680
(1990 Census)
Land Area: 48,843 Sq. Mi.
Capital: Raleigh

Cities and Towns

Aberdeen	C-6
Ahoskie	B-9
Albemarle	C-5
Andrews	C-1
Apex	C-7
Asheboro	C-6
Asheville	C-3
Bakersville	B-3
Bayboro	C-9
Beaufort	D-9
Belhaven	C-9
Benson	C-7
Blowing Rock	B-4
Boone	B-4
Brevard	C-2
Bryson City	C-2
Burgaw	D-8
Burlington	B-6
Burnsville	B-3
Camden	B-9
Canton	C-2
Carthage	C-6
Cary	B-7
Chadbourn	E-7
Chapel Hill	B-6
Charlotte	C-5
Cherokee	C-2
Cherryville	C-4
Clayton	C-7
Clinton	D-7
Columbia	B-9
Columbus	D-2
Concord	C-5
Currituck	A-10
Danbury	B-5
Dobson	B-5
Dunn	C-7
Durham	B-7
Eden	A-6
Edenton	B-9
Elizabeth City	B-9
Elizabethtown	D-7
Enfield	B-8
Erwin	C-7
Farmville	C-8
Fayetteville	D-7
Forest City	C-3
Franklin	C-2
Fuquay-Varina	C-7
Gastonia	C-4
Goldsboro	C-8
Graham	B-6
Granite Falls	B-4
Greensboro	B-6
Greenville	C-8
Halifax	B-8
Hamlet	D-6
Havelock	D-9
Hayesville	C-1
Henderson	B-7
Hendersonville	C-3
Hertford	B-9
Hickory	C-4
High Point	B-6
Hillsborough	B-6
Jackson	A-8
Jacksonville	D-8
Jefferson	B-4
Kannapolis	C-5
Kenansville	D-8
Kings Mtn.	C-4
Kinston	C-8
Laurinburg	D-6
Lenoir	B-4
Lexington	B-5
Lillington	C-7
Lincolnton	C-4
Longview	C-4
Louisburg	B-7
Lumberton	D-7
Madison	B-6
Manteo	B-10
Marion	C-3
Marshall	B-3
Mars Hill	B-3
Maxton	D-6
Maysville	D-8
Mocksville	B-5
Monroe	D-5
Mooresville	C-5
Morganton	B-3
Mt. Airy	A-5
Mt. Gilead	C-6
Mt. Holly	C-4
Mt. Olive	C-7
Murphy	C-1
Nashville	B-8
New Bern	C-9
Newland	B-3
Newton	C-4
Norwood	C-5
Oteen	C-3
Oxford	B-7
Pilot Mtn.	B-5
Pinehurst	C-6
Pineville	D-5
Pittsboro	C-6
Plymouth	B-9
Raeford	D-6
Raleigh	B-7
Red Sprs.	D-6
Reidsville	B-6
Roanoke Rapids	A-8
Robbinsville	C-1
Rockingham	D-6
Rocky Mt.	B-8
Rose Hill	D-7
Rowland	D-6
Roxboro	B-6
Rutherfordton	C-3
St. Paul	D-6
Salisbury	C-5
Sanford	C-6
Scotland Neck	B-8
Shelby	C-3
Smithfield	C-7
Snow Hill	C-8
Southern Pines	C-6
Southport	E-8
Sparta	A-4
Spring Hope	B-8
Spruce Pine	B-3
Statesville	B-5
Swannanoa	C-3
Swanquarter	C-9
Swansboro	D-9
Sylva	C-2
Tabor City	E-7
Tarboro	B-8
Taylorsville	B-4
Thomasville	B-5
Trenton	C-8
Troy	C-6
Valdese	C-4
Vanceboro	C-8
Wadesboro	D-5
Wake Forest	B-7
Wallace	D-8
Walnut Cove	B-5
Warrenton	B-7
Warsaw	D-7
Washington	C-9
Waynesville	C-2
Wentworth	B-6
Whiteville	E-7
Wilkesboro	B-4
Williamston	B-8
Wilmington	E-8
Wilson	C-8
Windsor	B-9
Winston-Salem	B-5
Winton	A-9
Yanceyville	B-6

SOUTH CAROLINA

Population: 3,505,707
(1990 Census)
Land Area: 30,207 Sq. Mi.
Capital: Columbia

Cities and Towns

Abbeville	E-3
Aiken	F-4
Allendale	G-4
Anderson	D-2
Andrews	F-6
Bamberg	F-4
Barnwell	F-4
Batesburg	E-4
Beaufort	G-5
Belton	D-3
Bennettsville	D-6
Bethune	D-5
Bishopville	E-5
Blackville	F-4
Branchville	F-5
Calhoun Falls	E-3
Camden	E-5
Charleston	G-6
Cheraw	D-6
Chester	D-4
Chesterfield	D-5
Clemson	D-2
Clinton	D-3
Columbia	E-4
Conway	E-6
Cowpens	D-4
Darlington	E-5
Denmark	F-4
Dillon	D-6
Easley	D-3
Edgefield	E-3
Elliott	E-5
Estill	G-4
Fairfax	G-4
Florence	E-6
Fountain Inn	D-3
Gaffney	D-4
Georgetown	F-6
Great Falls	D-4
Greeleyville	F-6
Greenville	D-3
Greenwood	E-3
Greer	D-3
Hampton	G-4
Hardeeville	H-4
Hartsville	E-5
Hemingway	E-6
Hodges	E-3
Holly Hill	F-5
Honea Path	D-3
Jacksonboro	G-5
Jamestown	F-6
Johnston	E-4
Jonesville	D-3
Kershaw	D-5
Kingstree	E-6
Lake City	E-6
Lancaster	D-5
Latta	D-6
Laurens	D-3
Lexington	E-4
Liberty	D-2
Loris	E-7
Lynchburg	E-5
McBee	D-5
McColl	D-6
McConnells	D-4
McCormick	E-3
Manning	F-5
Marion	E-6
Moncks Cor.	F-6
Mt. Pleasant	G-6
Murrells Inlet	F-6
Myrtle Beach	F-7
Newberry	E-4
Nichols	E-6
North	F-4
North Augusta	F-3
N. Charleston	G-6
Olanta	E-6
Orangeburg	F-5
Pageland	D-5
Pickens	D-2
Princeton	D-3
Rembert	E-5
Ridgeland	G-4
Rock Hill	D-4
St. George	F-5
St. Matthews	F-5
St. Stephen	F-6
Saluda	E-4
Seneca	D-2
Society Hill	D-6
Spartanburg	D-3
Summerton	F-5
Summerville	G-5
Sumter	E-5
Swansea	F-4
Timmonsville	E-5
Traveler's Rest	D-3
Union	D-4
Varnville	G-4
Walhalla	D-2
Waterboro	G-5
Ware Shoals	E-3
Westminster	D-2
Whitmire	D-4
Williston	F-4
Winnsboro	E-4
Woodruff	D-3
York	D-4

CEDAR ISLAND—OCRACOKE FERRY
Reservations requested from the point of departure within 30 days of departure.
Phone: Cedar Island (919) 225-3551
Ocracoke (919) 928-3841
Toll: Car and occupants—$10.00 one way

NORTH-SOUTH CAROLINA

Scale: 0 5 10 20 30 miles

© RAND McNALLY & CO.

PRINTED IN U.S.A.

93-1

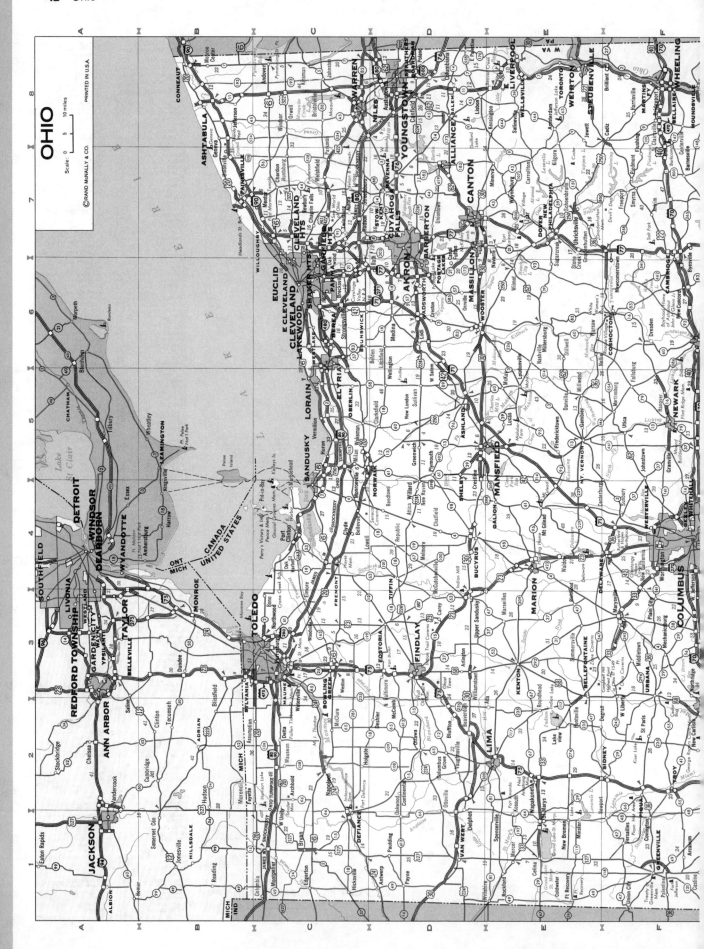

OHIO

Scale: 0 5 10 miles

©RAND McNALLY & CO.   PRINTED IN USA

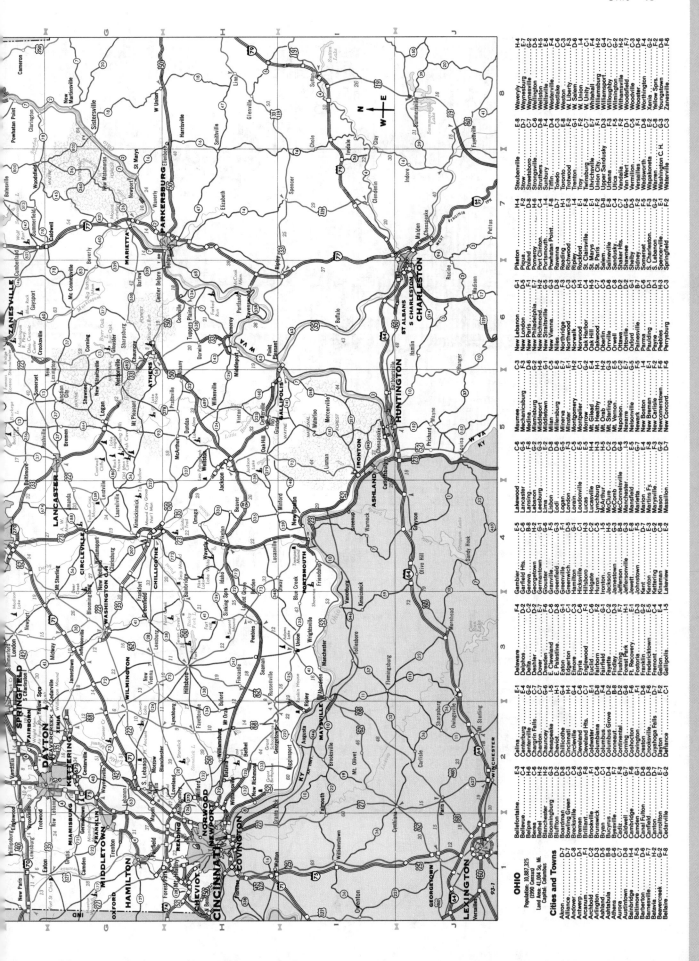

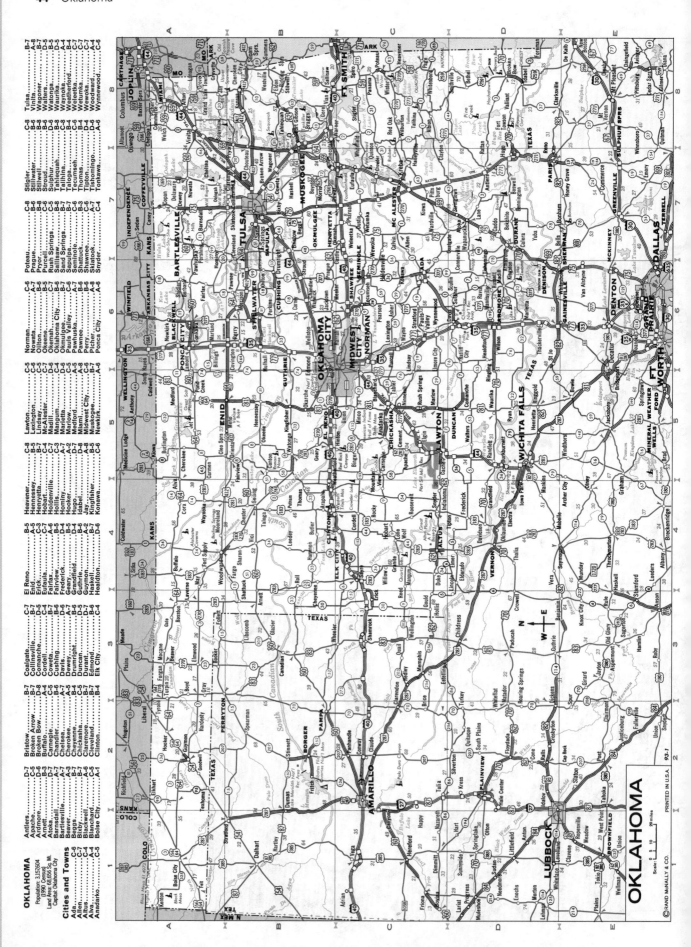

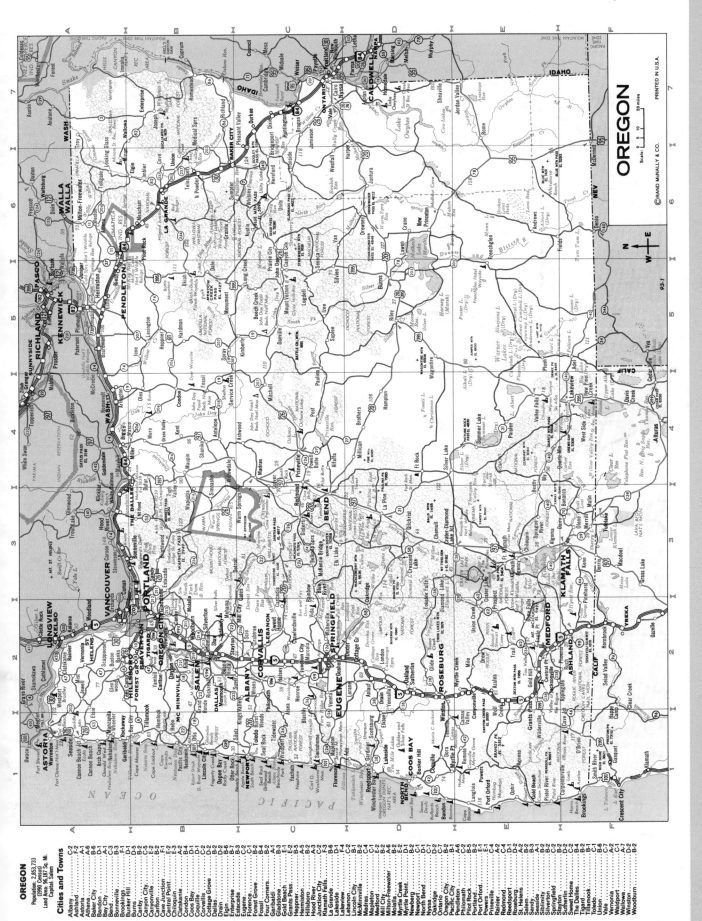

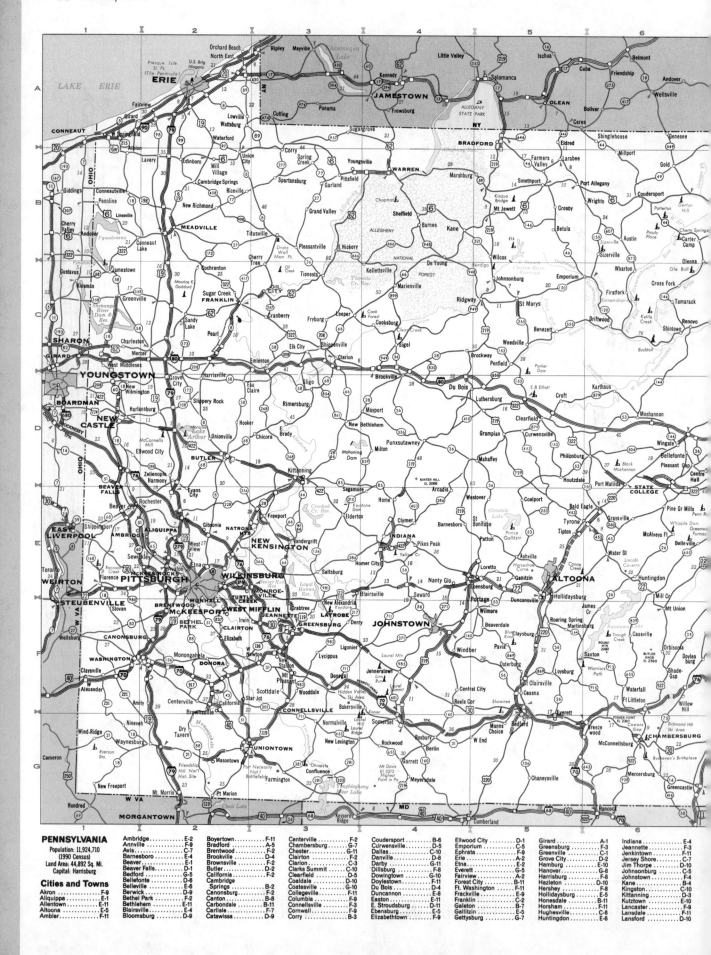

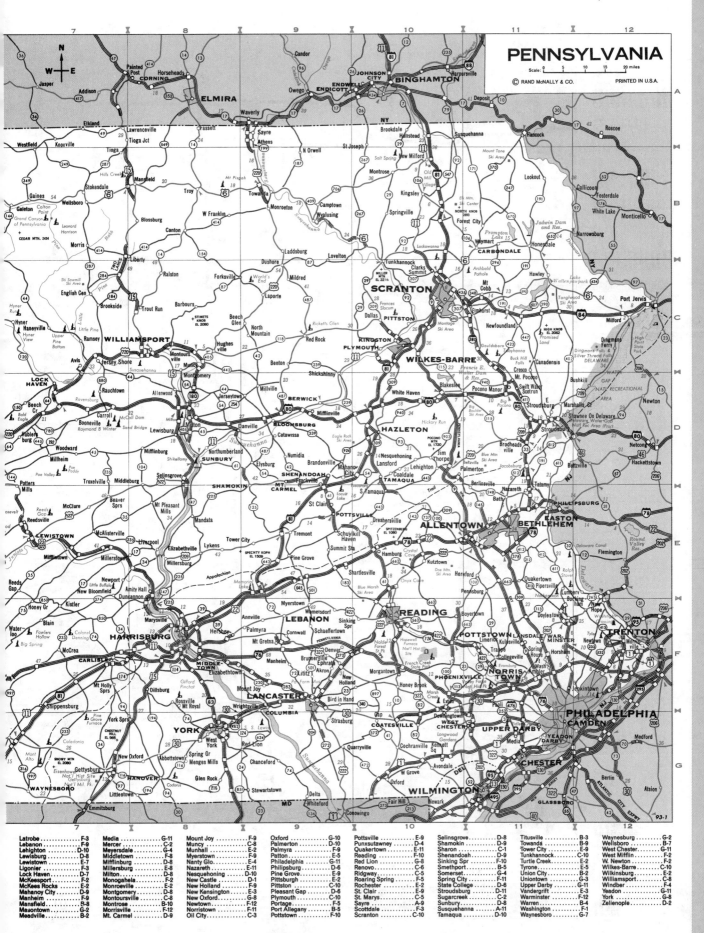

SOUTH DAKOTA

Population: 699,999 (1990 Census)
Land Area: 75,956 Sq. Mi.
Capital: Pierre

Cities and Towns

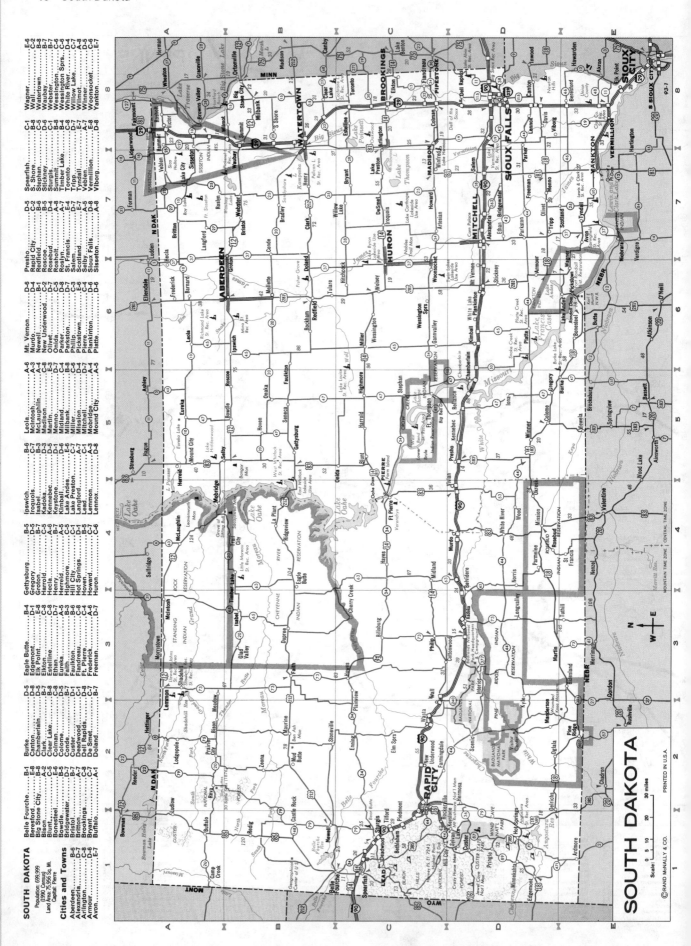

SOUTH DAKOTA

Scale:
0 5 10   20   30 30 miles

© RAND McNALLY & CO.

PRINTED IN U.S.A.

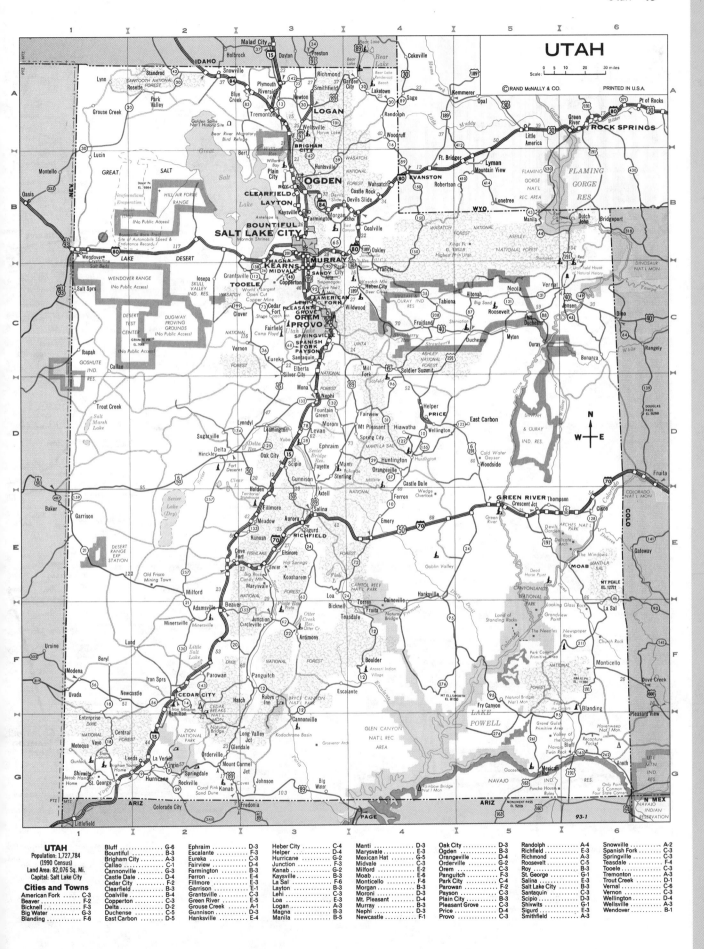

UTAH

Scale:
©RAND McNALLY & CO.   PRINTED IN U.S.A.

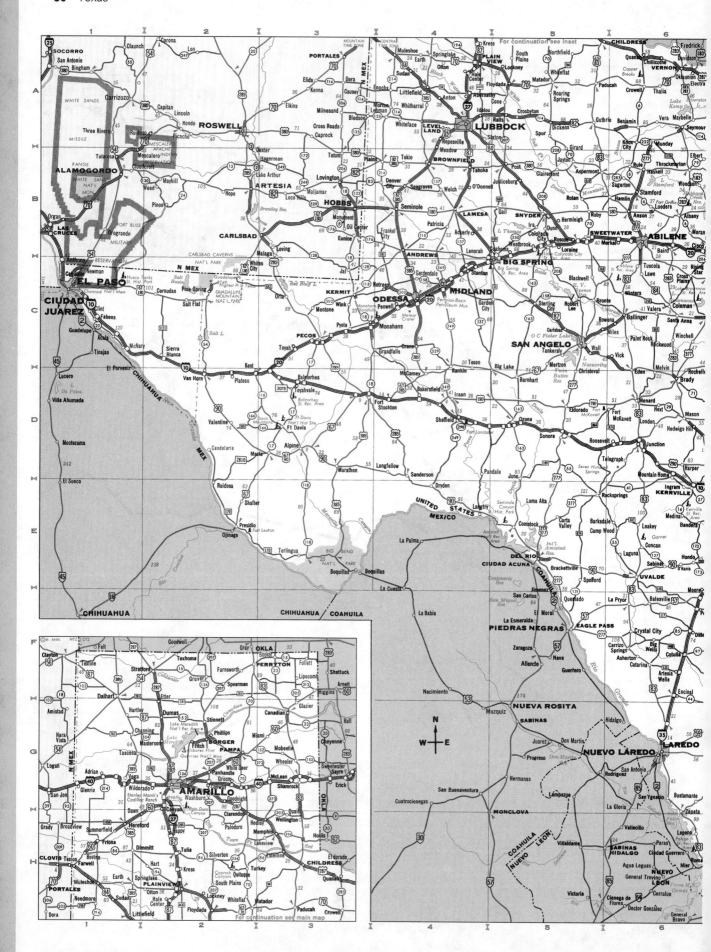

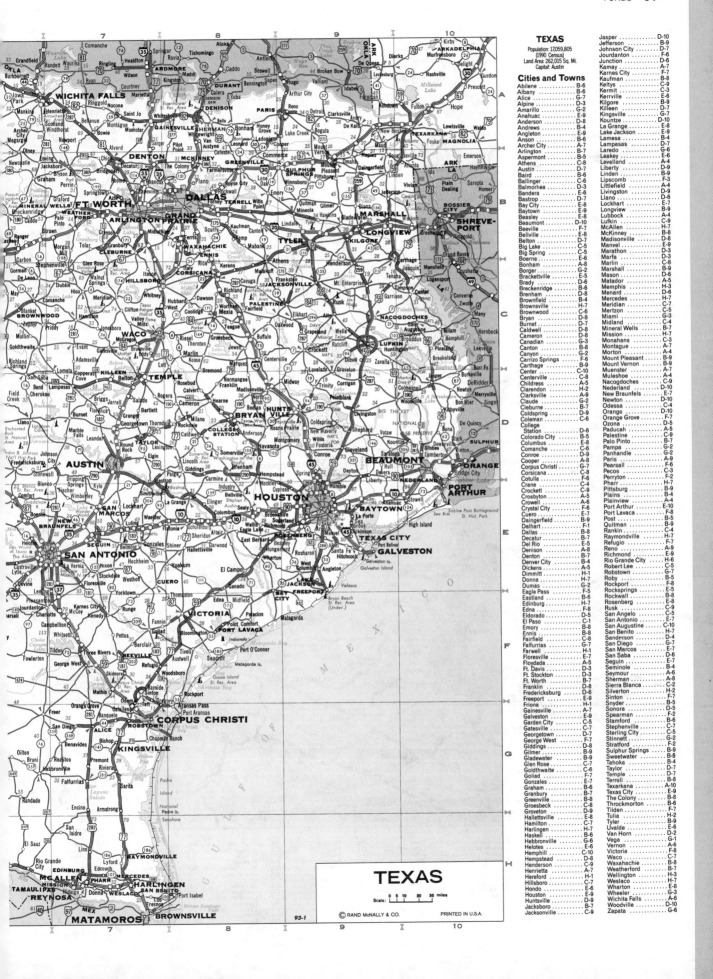

TEXAS

Population: 17,059,805
(1990 Census)
Land Area: 262,015 Sq. Mi.
Capital: Austin

Cities and Towns

Abilene	B-6
Albany	B-6
Alice	G-7
Alpine	D-3
Amarillo	G-2
Anahuac	E-9
Anderson	D-8
Andrews	B-4
Angleton	E-9
Anson	B-6
Archer City	A-7
Arlington	B-5
Aspermont	B-5
Athens	C-8
Austin	D-7
Baird	B-6
Ballinger	C-6
Balmorhea	D-3
Bandera	D-6
Bastrop	D-7
Bay City	E-8
Baytown	E-8
Beasley	E-8
Beaumont	D-10
Beeville	F-7
Bellville	E-8
Belton	D-7
Big Lake	C-5
Big Spring	C-5
Boerne	E-6
Bonham	A-8
Borger	E-5
Brackettville	D-6
Brady	D-6
Breckenridge	B-6
Brenham	D-8
Brownfield	B-4
Brownsville	H-7
Brownwood	C-6
Bryan	D-7
Burnet	D-7
Caldwell	D-8
Cameron	D-8
Canadian	G-3
Canton	B-8
Canyon	G-2
Carrizo Springs	F-6
Carthage	C-10
Center	C-10
Centerville	C-8
Childress	A-5
Clarendon	G-2
Clarksville	H-2
Claude	G-2
Cleburne	B-7
Coldspring	D-9
Coleman	C-6
College Station	D-8
Colorado City	C-5
Columbus	E-8
Comanche	C-6
Conroe	D-9
Cooper	A-8
Corpus Christi	G-8
Corsicana	C-8
Cotulla	F-6
Crane	C-4
Crockett	C-9
Crosbyton	A-5
Crowell	A-6
Crystal City	E-7
Cuero	E-7
Daingerfield	B-9
Dalhart	F-1
Dallas	B-8
Decatur	B-7
Del Rio	E-5
Denison	B-7
Denton	B-7
Denver City	B-4
Dickens	A-5
Dimmitt	H-1
Donna	H-7
Dumas	F-5
Eagle Pass	E-6
Eastland	B-6
Edinburg	H-7
Edna	F-8
Eldorado	C-5
El Paso	C-1
Emory	B-8
Ennis	B-8
Fairfield	C-8
Falfurrias	H-1
Farwell	H-1
Floresville	E-7
Floydada	A-5
Ft. Davis	D-3
Ft. Stockton	D-3
Ft. Worth	B-7
Franklin	C-8
Fredericksburg	D-6
Freeport	F-9
Friona	H-1
Gainesville	A-7
Galveston	E-9
Garden City	C-5
Gatesville	C-7
Georgetown	D-7
George West	F-7
Giddings	D-8
Gilmer	B-9
Gladewater	B-9
Glen Rose	C-7
Goldthwaite	C-6
Goliad	F-7
Gonzales	E-7
Graham	B-6
Granbury	B-7
Greenville	B-8
Groesbeck	C-8
Groveton	D-9
Hallettsville	E-8
Hamilton	C-7
Harlingen	H-7
Haskell	B-6
Hebbronville	G-6
Helotes	E-6
Hemphill	C-10
Hempstead	D-8
Henderson	C-9
Henrietta	A-7
Hereford	H-1
Hillsboro	C-7
Hondo	E-6
Houston	E-9
Huntsville	D-9
Jacksboro	B-7
Jacksonville	C-9

Jasper	D-10
Jefferson	B-9
Johnson City	D-7
Jourdanton	F-6
Junction	D-6
Kamay	A-7
Karnes City	F-7
Kaufman	B-8
Keltys	C-9
Kermit	C-3
Kerrville	E-6
Kilgore	B-9
Killeen	D-7
Kingsville	G-7
Kountze	D-10
La Grange	E-8
Lake Jackson	E-9
Lamesa	B-4
Lampasas	C-6
Laredo	G-6
Leakey	E-6
Levelland	A-4
Liberty	D-9
Linden	B-9
Lipscomb	F-3
Littlefield	A-4
Livingston	D-9
Llano	D-6
Lockhart	E-7
Longview	B-9
Lubbock	A-4
Lufkin	C-9
Luling	E-7
McAllen	H-7
McKinney	B-8
Madisonville	D-8
Manvel	E-9
Marathon	D-3
Marfa	D-3
Marlin	C-8
Marshall	B-9
Mason	D-6
Matador	A-5
Memphis	H-3
Menard	D-6
Mercedes	H-7
Meridian	C-7
Mertzon	C-5
Miami	G-3
Midland	C-4
Mineral Wells	B-7
Mission	H-7
Monahans	C-3
Montague	A-7
Morton	A-4
Mount Pleasant	B-9
Mount Vernon	B-9
Muenster	A-7
Muleshoe	A-4
Nacogdoches	C-9
Nederland	D-10
New Braunfels	E-7
Newton	D-10
Odessa	C-4
Orange	D-10
Orange Grove	G-7
Ozona	D-5
Paducah	A-5
Palestine	C-8
Palo Pinto	B-7
Pampa	G-2
Panhandle	G-2
Paris	A-8
Pearsall	F-6
Pecos	C-3
Perryton	F-2
Pharr	H-7
Pittsburg	B-9
Plains	B-4
Plainview	A-4
Port Arthur	E-10
Port Lavaca	F-8
Post	B-5
Quitman	B-9
Rankin	C-4
Raymondville	H-7
Refugio	F-7
Reno	E-9
Richmond	E-9
Rio Grande City	H-6
Robert Lee	C-5
Robstown	G-7
Roby	B-5
Rockport	F-8
Rocksprings	E-5
Rockwall	B-8
Rosenberg	E-8
Rusk	C-9
San Angelo	C-5
San Antonio	E-7
San Augustine	C-10
San Benito	H-7
Sanderson	D-4
San Diego	G-7
San Marcos	E-7
San Saba	D-6
Seguin	E-7
Seminole	B-4
Seymour	A-6
Sherman	A-8
Sierra Blanca	C-2
Silverton	H-2
Sinton	F-7
Snyder	B-5
Sonora	D-5
Spearman	F-2
Stamford	B-6
Stephenville	C-7
Sterling City	C-5
Stinnett	G-2
Stratford	F-2
Sulphur Springs	B-8
Sweetwater	B-5
Tahoka	B-4
Taylor	D-7
Temple	D-7
Terrell	B-8
Texarkana	A-10
Texas City	E-9
The Colony	B-8
Throckmorton	B-6
Tilden	F-7
Tulia	H-2
Tyler	B-9
Uvalde	E-6
Van Horn	D-2
Vega	G-1
Vernon	A-6
Victoria	F-8
Waco	C-7
Waxahachie	B-8
Weatherford	B-7
Wellington	H-3
Weslaco	H-7
Wharton	E-8
Wheeler	G-3
Wichita Falls	A-7
Woodville	D-10
Zapata	G-6

TEXAS

Scale: 0 5 10 20 30 miles

93-1   © RAND MCNALLY & CO.   PRINTED IN U.S.A.

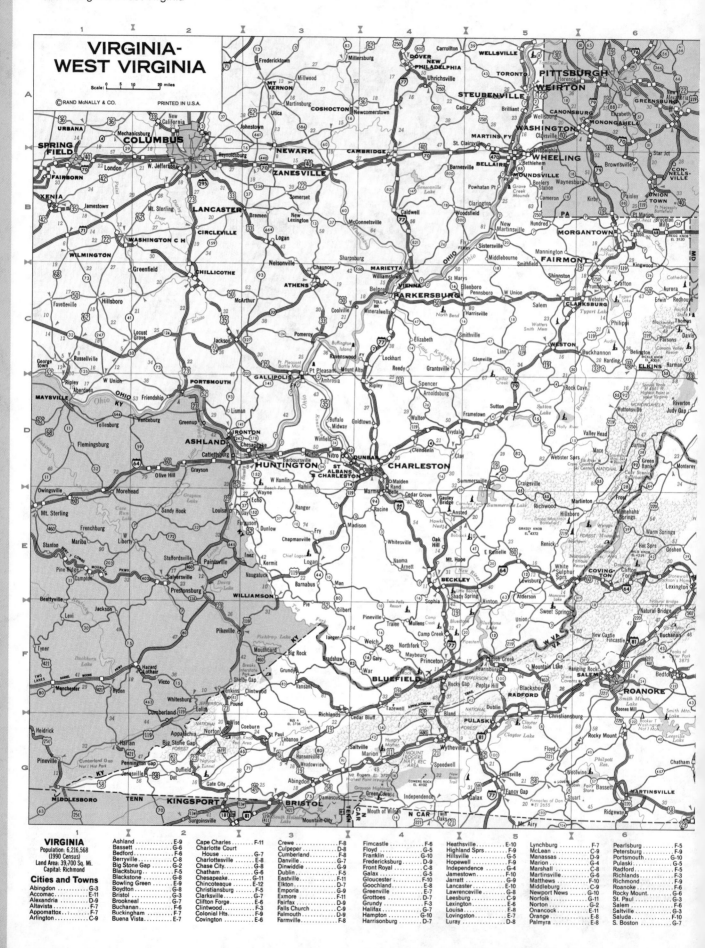

VIRGINIA-WEST VIRGINIA

Scale: 0 5 10   20 miles

©RAND McNALLY & CO.   PRINTED IN U.S.A.

VIRGINIA

Population: 6,216,568
(1990 Census)
Land Area: 39,700 Sq. Mi.
Capital: Richmond

Cities and Towns

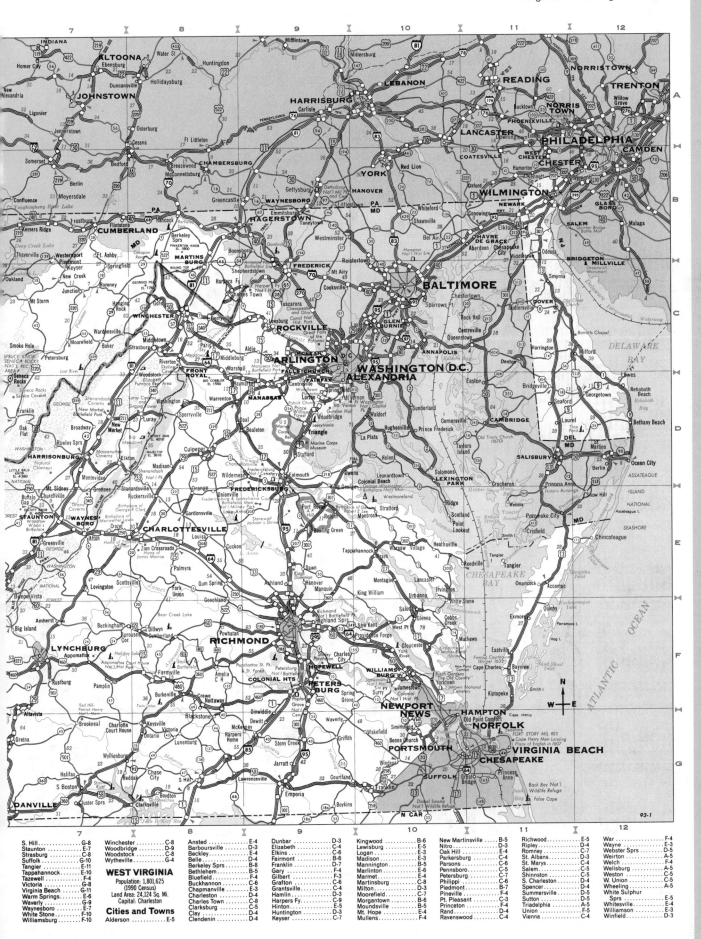

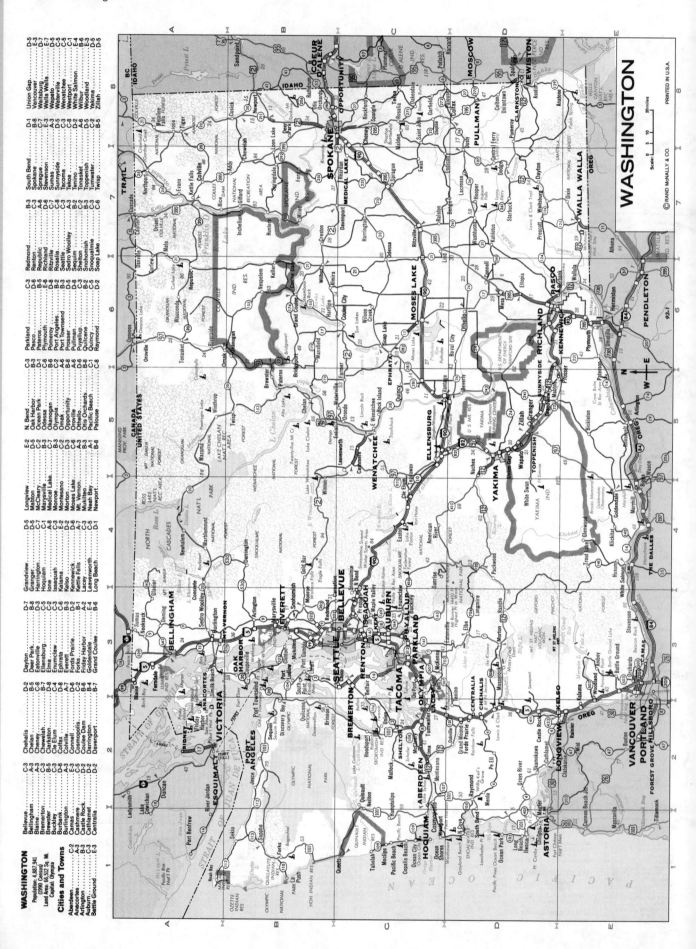

WASHINGTON

© RAND McNALLY & CO.

PRINTED IN U.S.A.

93-1

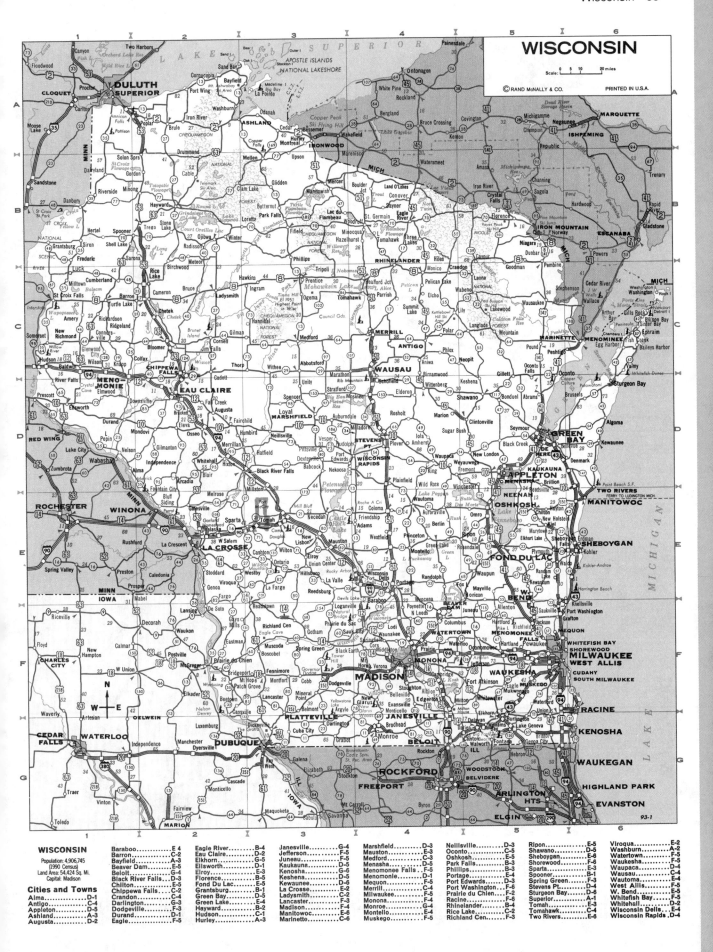

93-1

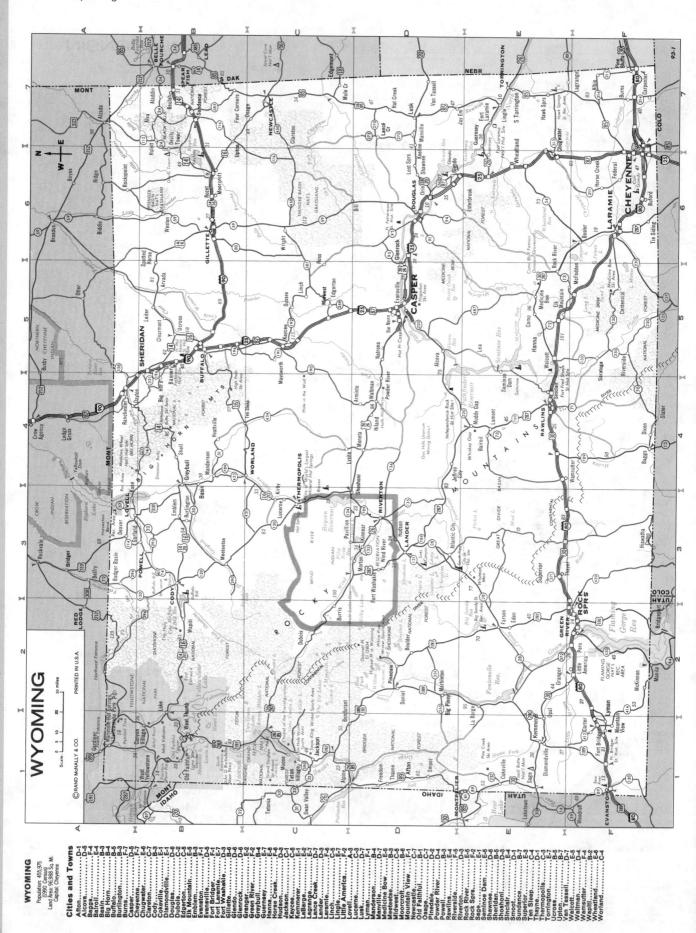

WYOMING

Population: 455,975
(1990 Census)
Land Area: 96,988 Sq. Mi.
Capital: Cheyenne

©RAND McNALLY & CO.

PRINTED IN U.S.A.

Scale
0 5 10   20   30 miles

Cities and Towns

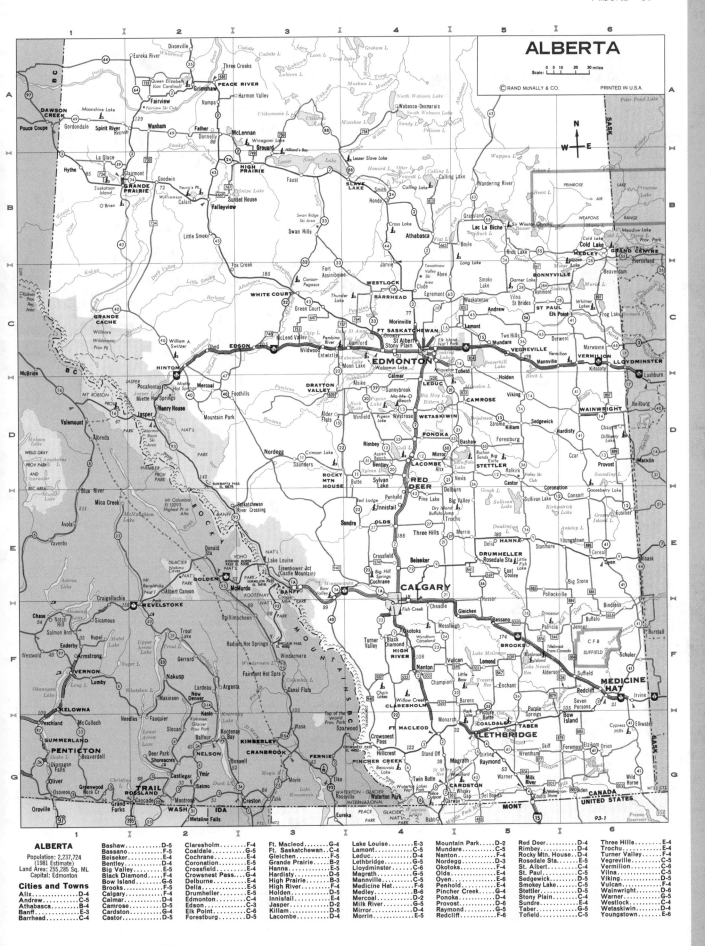

ALBERTA

Scale: 0 5 10 20 30 miles

© RAND McNALLY & CO.   PRINTED IN U.S.A.

ALBERTA

Population: 2,237,724
(1981 Estimate)
Land Area: 255,285 Sq. Mi.
Capital: Edmonton

Cities and Towns

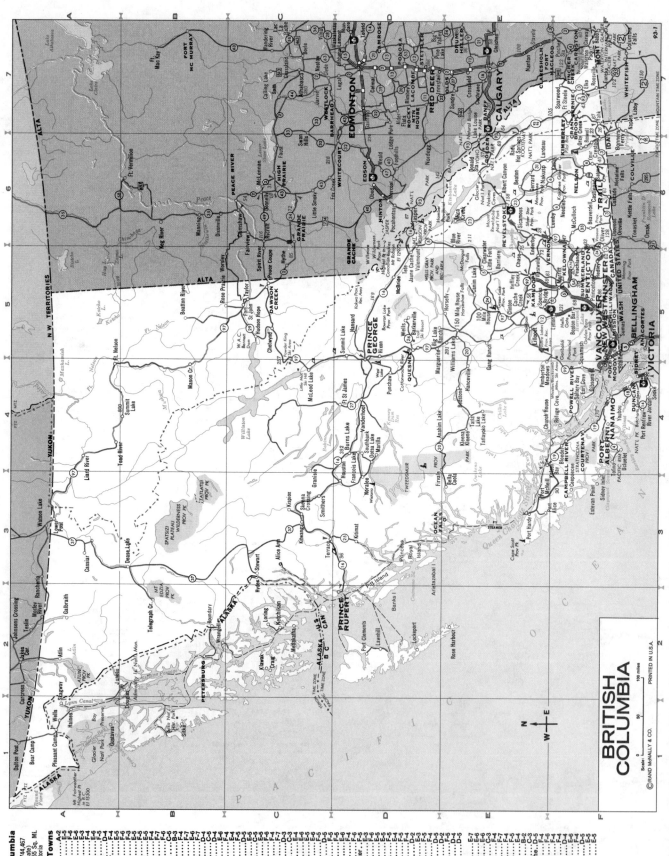

British Columbia

Population: 2,744,467
(1981 Estimate)
Land Area: 366,255 Sq. Mi.
Capital: Victoria

Cities and Towns

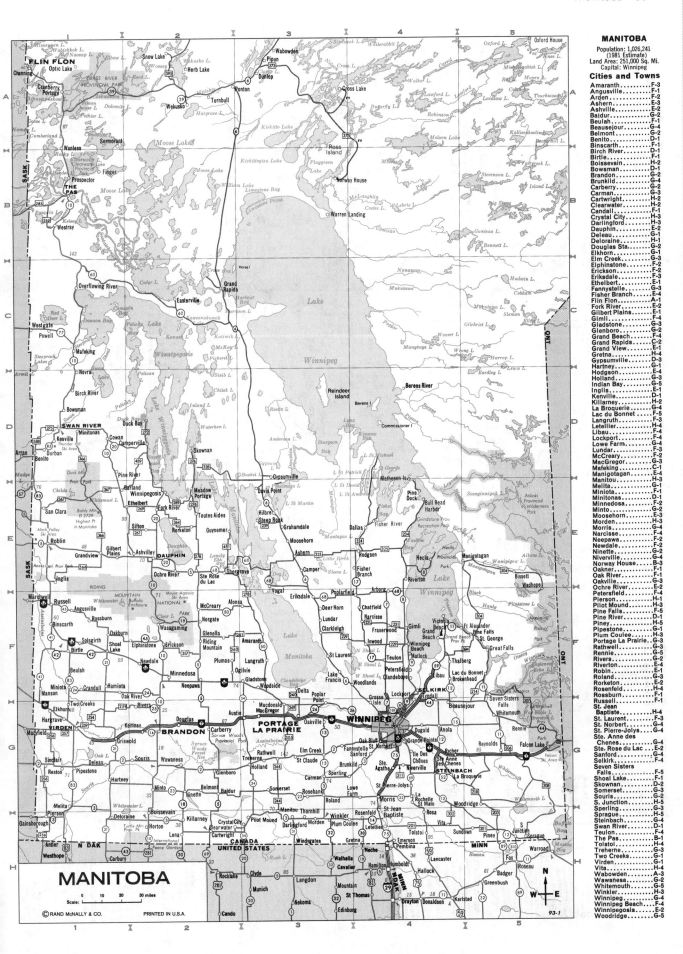

MANITOBA

Population: 1,026,241
(1981 Estimate)
Land Area: 251,000 Sq. Mi.
Capital: Winnipeg

Cities and Towns

Amaranth.........F-3
Angusville.......F-1
Arden............F-2
Ashern...........E-3
Ashville.........E-2
Baldur...........G-2
Beulah...........F-1
Beausejour.......G-4
Belmont..........G-2
Benito...........D-1
Binscarth........F-1
Birch River......D-1
Birtle...........F-1
Boissevain.......H-2
Bowsman..........D-1
Brandon..........G-2
Brunkild.........G-2
Carberry.........G-3
Carman...........G-3
Cartwright.......H-2
Clearwater.......H-2
Candall..........G-2
Crystal City.....H-3
Darlingford......H-3
Dauphin..........E-2
Deleau...........G-1
Deloraine........H-1
Douglas Sta......G-2
Elkhorn..........G-1
Elm Creek........G-3
Elphinstone......F-2
Erickson.........F-2
Eriksdale........F-3
Ethelbert........E-1
Fannystelle......G-3
Fisher Branch....E-4
Flin Flon........A-1
Fork River.......E-2
Gilbert Plains...F-1
Gimli............F-4
Gladstone........G-3
Glenboro.........G-2
Grand Beach......F-4
Grand Rapids.....C-2
Grand View.......E-1
Gretna...........H-3
Gypsumville......D-3
Hartney..........G-1
Hodgson..........E-4
Holland..........G-2
Indian Bay.......G-5
Inglis...........E-1
Kenville.........E-1
Killarney........H-2
La Broquerie.....G-4
Lac du Bonnet....F-5
Langruth.........F-3
Letellier........H-4
Libau............F-4
Lockport.........G-4
Lowe Farm........G-3
Lundar...........F-3
McCreary.........F-2
MacGregor........G-3
Mafeking.........C-1
Manigotagan......F-4
Manitou..........H-3
Melita...........G-1
Minitota.........G-1
Minitonas........D-1
Minnedosa........G-2
Minto............G-2
Moosehorn........E-3
Morden...........H-3
Morris...........G-4
Narcisse.........F-4
Neepawa..........F-2
Newdale..........G-2
Ninette..........G-2
Niverville.......G-4
Norway House.....B-3
Oakner...........F-1
Oak River........F-1
Oakville.........G-3
Ochre River......E-2
Petersfield......F-4
Pierson..........H-1
Pilot Mound......H-3
Pine Falls.......F-5
Pine River.......D-1
Piney............H-5
Pipestone........G-1
Plum Coulee......H-3
Portage La Prairie.G-3
Rathwell.........G-3
Rennie...........G-5
Rivers...........G-2
Riverton.........E-4
Robin............E-1
Roland...........G-4
Rorketon.........E-2
Rosenfeld........H-4
Rossburn.........F-1
Russell..........F-1
St. Jean
  Baptiste.......H-4
St. Laurent......F-3
St. Norbert......G-4
St. Pierre-Jolys.G-4
Ste. Anne des
  Chenes.........G-4
Ste. Rose du Lac.E-2
Sanford..........G-4
Selkirk..........F-4
Seven Sisters
  Falls..........F-5
Shoal Lake.......F-1
Skownan..........D-2
Somerset.........G-3
Souris...........G-2
S. Junction......H-5
Sperling.........G-3
Sprague..........H-5
Steinbach........G-4
Swan River.......D-1
Teulon...........F-4
The Pas..........B-1
Tolstoi..........H-4
Treherne.........G-3
Two Creeks.......G-1
Virden...........G-1
Vita.............H-4
Wabowden.........A-3
Wawanesa.........G-2
Whitemouth.......G-4
Winkler..........H-3
Winnipeg.........G-4
Winnipeg Beach...F-4
Winnipegosis.....E-2
Woodridge........G-5

93-1

NEW BRUNSWICK

Population: 696,403
(1981 Estimate)
Land Area: 28,354 Sq. Mi.
Capital: Fredericton

Cities and Towns

Bathurst............B-3
Bouctouche..........C-4
Campbellton.........B-3
Chatham.............C-3
Chipman.............D-3
Dalhousie...........B-3
Edmundston..........B-2
Fredericton.........D-3
Grand Falls.........C-2
Kedgwick............B-2
Minto...............D-3
Moncton.............D-4
Newcastle...........C-3
Plaster Rock........C-2
Richibucto..........C-4
Saint John..........E-3
St. Léonard.........C-2
St. Quentin.........B-2
St. Stephen.........E-2
Sussex..............D-3

NOVA SCOTIA

Population: 847,442
(1981 Estimate)
Land Area: 21,425 Sq. Mi.
Capital: Halifax

Cities and Towns

Advocate Hbr........E-4
Amherst.............D-4
Annapolis Royal.....E-3
Antigonish..........D-6
Bass River..........D-5
Bridgetown..........E-3
Bridgewater.........E-4
Canso...............D-7
Chester.............E-4
Cheticamp...........C-6
Darthmouth..........E-5
Digby...............E-3
Glace Bay...........D-7
Halifax.............E-5
Ingonish............C-7
Kentville...........E-4
Liverpool...........F-4
Lockeport...........F-4
Lunenburg...........E-4
Mahone Bay..........E-4
Middleton...........E-4
New Germany.........E-4
New Glasgow.........D-5
New Waterford.......C-7
North Sydney........D-7
Perrsboro...........D-4
Pictou..............D-5
Shelburne...........F-3
Springhill..........D-4
Sydney..............D-7
Sydney Mines........C-7
Tetamagouche........D-5
Truro...............D-5
Wolfville...........E-4
Yarmouth............F-3

PRINCE EDWARD ISLAND

Population: 122,506
(1981 Estimate)
Land Area: 2,184 Sq. Mi.
Capital: Charlottetown

Cities and Towns

Cardigan............D-5
Charlottetown.......D-5
Georgetown..........D-5
Murray Harbour......D-5
Port Borden.........D-5
St. Peters Bay......C-5
Summerside..........C-4
Tignish.............C-4

NEWFOUNDLAND

Population: 567,681
(1981 Estimate)
Land Area: 156,185 Sq. Mi.
Capital: St. John's

Cities and Towns

Bishop's Falls......B-6
Bonavista...........B-7
Corner Brook........B-6
Grand Falls.........B-6
St. John's..........B-7

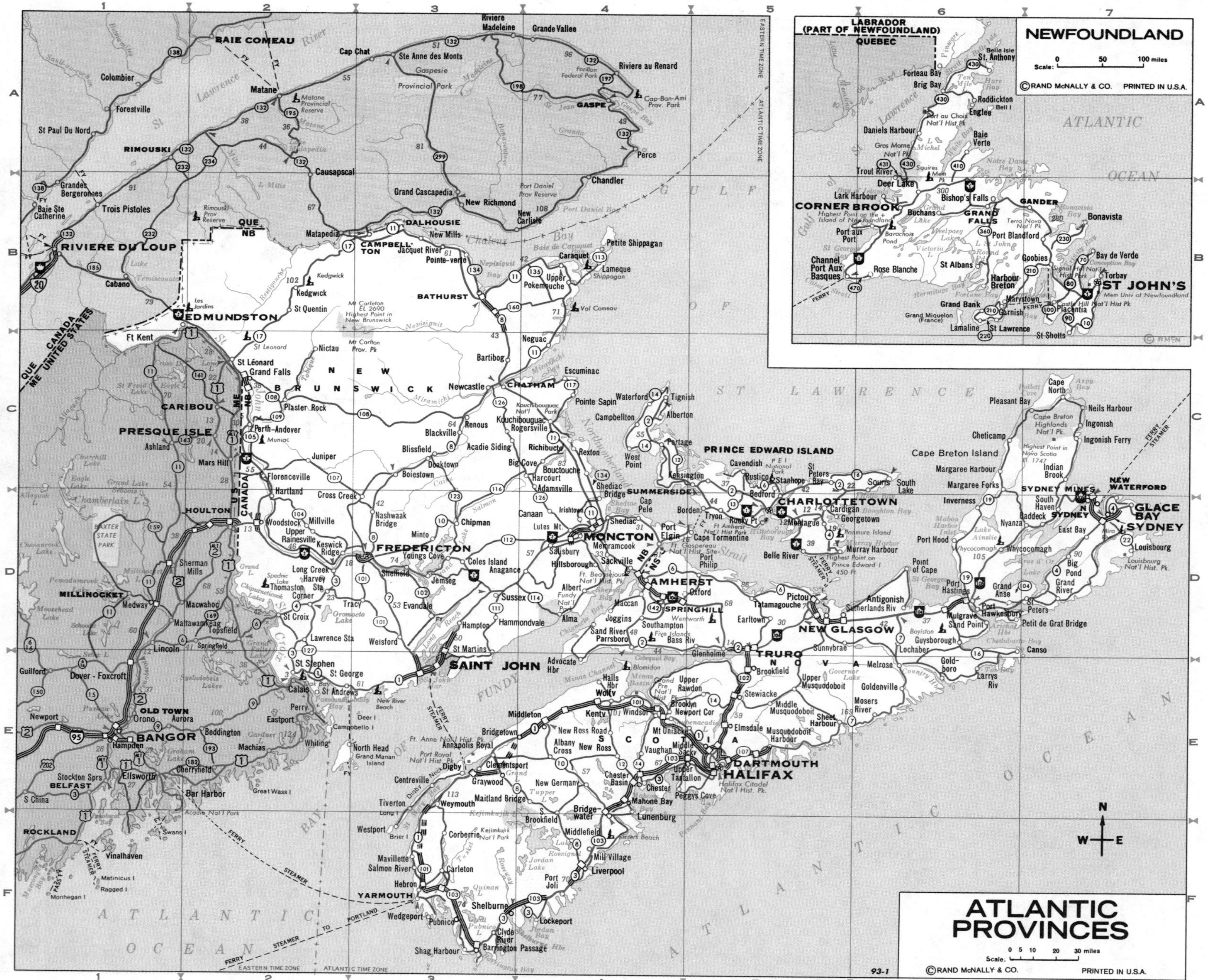
ATLANTIC PROVINCES
Scale. 0 5 10 20 30 miles
©RAND McNALLY & CO.
PRINTED IN U.S.A.
93-1
NEWFOUNDLAND
Scale: 0 50 100 miles
©RAND McNALLY & CO. PRINTED IN U.S.A.
LABRADOR (PART OF NEWFOUNDLAND)
QUEBEC
EASTERN TIME ZONE
ATLANTIC TIME ZONE

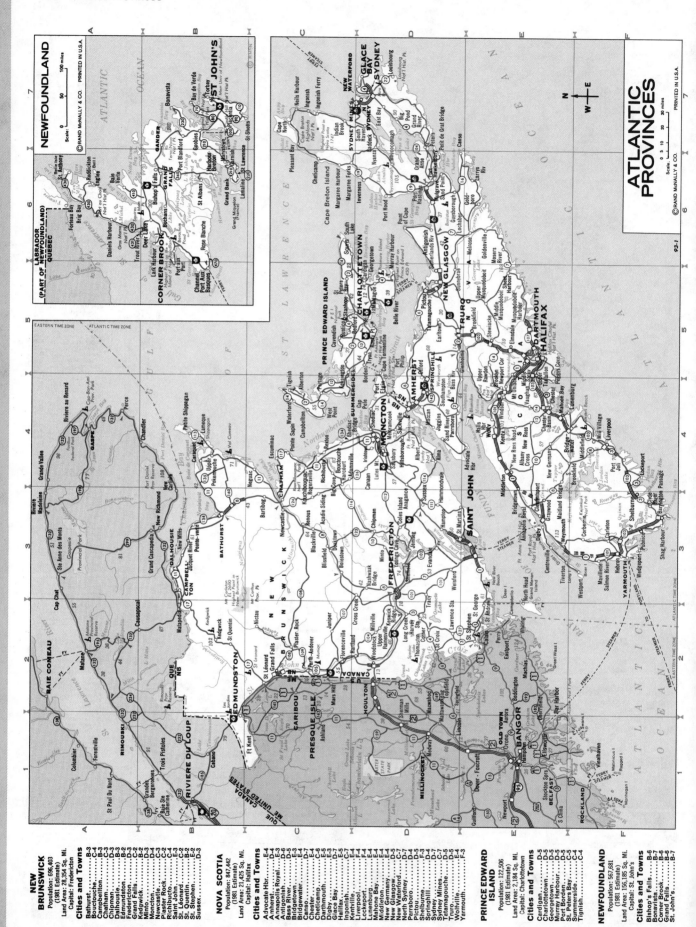

ONTARIO

Scale:
0   10   20   30 miles

© RAND McNALLY & CO.
PRINTED IN U.S.A.

For continuation see inset

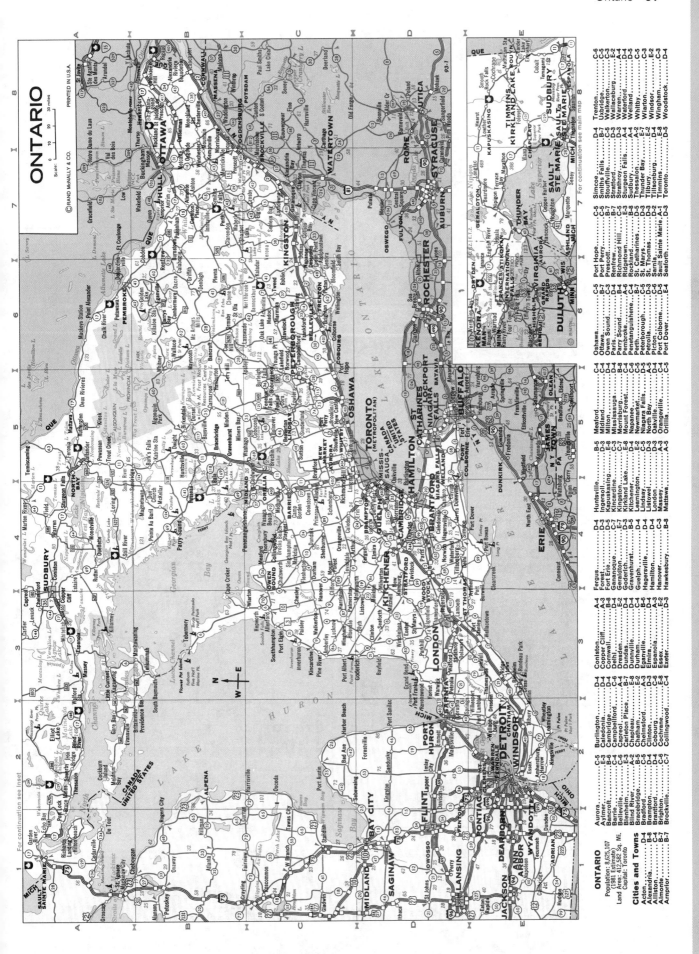

ONTARIO

Population: 8,625,107
(1981 Estimate)
Land Area: 412,582 Sq. Mi.
Capital: Toronto

Cities and Towns

Acton	B-8
Alexandria	C-4
Alliston	B-8
Almonte	B-7
Arnprior	B-7
Aurora	C-5
Aylmer	E-3
Bancroft	B-6
Barrie	C-6
Belleville	C-6
Blenheim	E-3
Blind River	A-2
Bracebridge	B-5
Brampton	D-4
Brantford	D-4
Brighton	C-6
Brockville	B-7
Burlington	C-5
Caledonia	E-3
Cambridge	E-6
Campbellford	C-6
Capreol	D-4
Carleton Place	B-7
Chapleau	A-2
Chatham	C-4
Chelmsford	D-4
Clinton	D-4
Cobourg	D-4
Cochrane	C-7
Coniston	D-4
Copper Cliff	D-4
Cornwall	B-8
Delhi	E-3
Dresden	C-7
Dundas	D-4
Dunnville	E-4
Durham	C-5
Espanola	D-3
Essex	C-6
Exeter	E-8
Fergus	A-3
Forest	B-8
Fort Erie	B-8
Fort Frances	E-2
Geraldton	E-7
Goderich	D-4
Gravenhurst	B-5
Guelph	A-6
Hagersville	E-2
Hamilton	D-3
Hanover	A-3
Hawkesbury	B-8
Huntsville	D-4
Ingersoll	D-5
Kapuskasing	C-7
Kincardine	E-7
Kingston	C-7
Kirkland Lake	B-5
Kitchener	E-8
Lindsay	D-4
Listowel	A-3
London	D-4
Longlac	C-3
Massey	E-2
Mattawa	B-8
Meaford	B-5
Midland	E-8
Milton	C-5
Mississauga	C-3
Morrisburg	C-7
Mount Forest	D-4
Napanee	B-5
New Liskeard	E-8
Niagara Falls	C-5
North Bay	C-6
Oakville	D-3
Orangeville	A-5
Orillia	A-5
Oshawa	C-4
Ottawa	D-4
Owen Sound	D-4
Paris	B-8
Parry Sound	C-3
Pembroke	B-4
Penetanguishene	A-6
Perth	C-5
Peterborough	D-3
Petrolia	D-3
Picton	D-2
Port Colborne	A-3
Port Dover	A-5
Port Hope	C-5
Port Perry	C-5
Prescott	C-3
Renfrew	D-4
Richmond Hill	B-4
Ridgetown	A-6
Rockland	B-8
St. Catharines	C-5
St. Marys	C-6
St. Thomas	A-5
Sarnia	D-3
Sault Sainte Marie	A-1
Seaforth	D-5
Simcoe	C-6
Smiths Falls	B-7
Stouffville	C-5
Stratford	D-4
Strathroy	A-4
Sturgeon Falls	E-3
Sudbury	D-4
Thunder Bay	E-2
Tilbury	D-3
Tillsonburg	E-2
Timmins	E-3
Toronto	D-4
Trenton	C-6
Uxbridge	B-7
Vankleek Hill	C-5
Walkerton	C-3
Wallaceburg	E-2
Warren	D-4
Waterford	A-4
Welland	A-7
Whitby	E-2
Wiarton	D-4
Windsor	E-2
Wingham	E-3
Woodstock	D-4

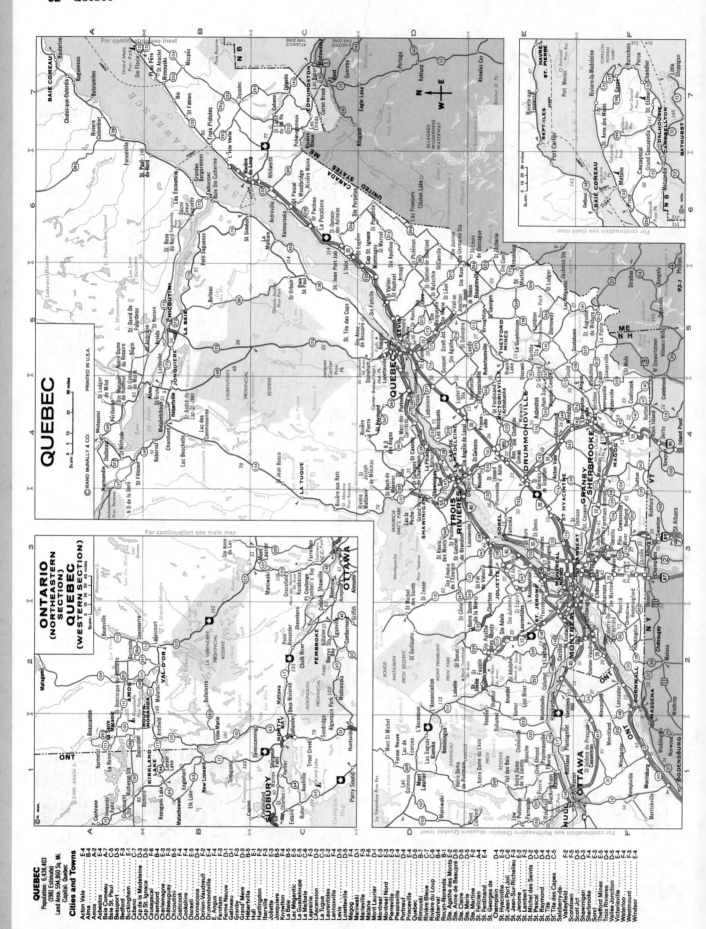

QUEBEC

© RAND McNALLY & CO.

PRINTED IN U.S.A.

Scale 0 10 20 30 miles

ONTARIO (NORTHEASTERN SECTION) QUEBEC (WESTERN SECTION)

Scale 0 10 20 30 40 miles

For continuation see inset

For continuation see main map

For continuation see Northeastern Ontario–Western Quebec Inset

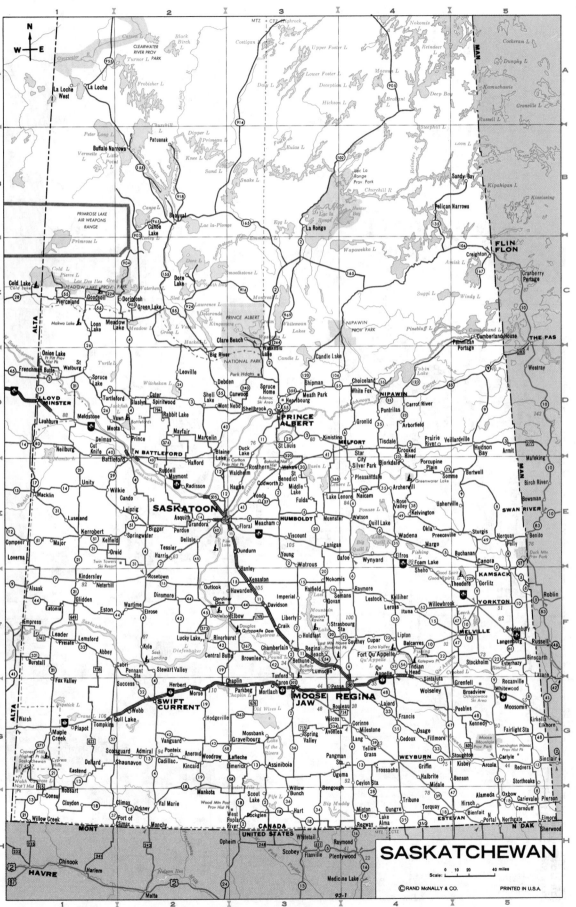

SASKATCHEWAN
Population: 968,313
(1981 Estimate)
Land Area: 251,700 Sq. Mi.
Capital: Regina

Cities and Towns

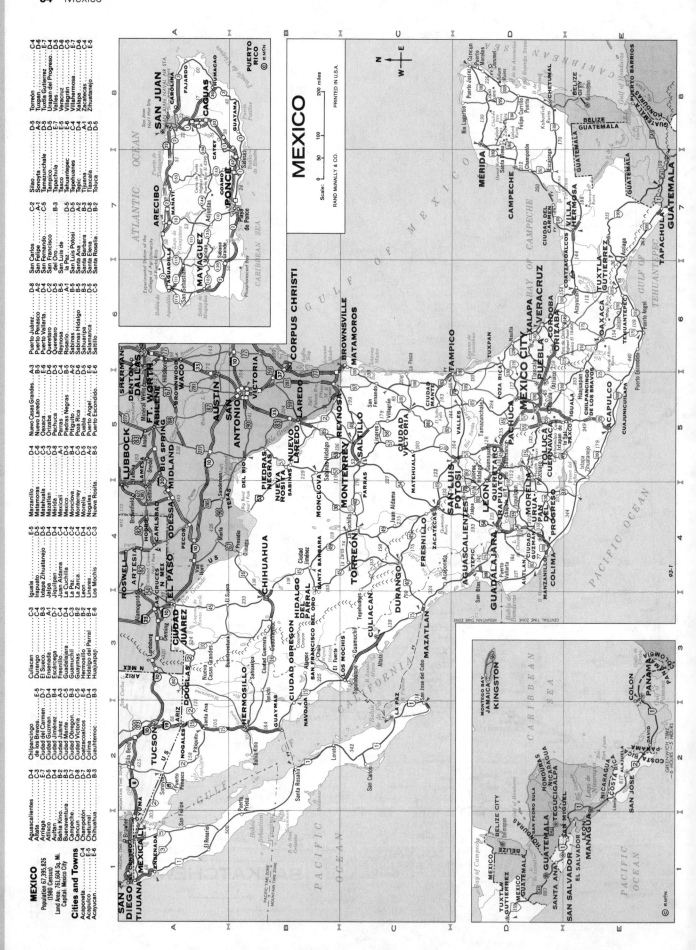

Major Cities Guide

Use the information in this section to acquaint yourself with 70 major cities, including principal United States cities, Mexico City, and the Canadian cities of Calgary, Edmonton, Montréal, Toronto, and Vancouver. The following information is provided:

Population Statistics. Populations of cities, figure followed by (1990C), and their metropolitan areas (figure preceded by an asterisk) are the census figures for the year cited. The U.S. metropolitan areas have been defined by the Bureau of the Census and the U.S. Office of Management and Budget to include one or more major cities and their counties. A metropolitan area may also include additional counties that have strong economic and social ties to the central county or counties.

Weather Information. Altitude figures, and average January and July temperatures, provide weather information in brief for each city.

Telephone Numbers/Time Zones. Telephone area codes and time zone information help orient you to each new location. Local time and weather numbers are provided where applicable.

Airport Transportation/Maps. Flyers will find useful information in each city listing under Airport Transportation: the distance from each city's airport(s) to downtown, and the type of transportation service that is available.

In addition, detailed airport maps for 27 of the largest cities are included in this section.

Hotels and Restaurants. A listing of selected hotels/motels and restaurants is provided for each city; establishments chosen for their broad appeal to the traveling public.

Attractions. A selected list of important local attractions, their addresses and telephone numbers, is provided for each city in the guide.

City Maps. Get around each city easily by using the large-scale maps for each city to identify streets, suburbs, freeways, hospitals, airports, points of interest, and more.

Tourist Information Sources. Each city listing also provides information sources for that city. Addresses and/or phone numbers are given for local convention and visitors bureaus, offices of tourism, and chambers of commerce.

Whether you call or write ahead for information, or stop in once you're in town, these offices can provide special information about their city that will enhance your visit and make the most of your sightseeing time.

Contents/Cities and Airports

Legend — City Maps

Roads and Related Symbols

Free Limited-Access Highways

Under Construction

Toll Limited-Access Highways

Under Construction

Other Four-Lane Divided Highways

Principal Highways

Other Through Highways

Other Roads (conditions vary — local inquiry suggested)

Interchanges and Numbers

Service Areas; Toll Booths

Information Center

Red accumulated mileages between red dots
Other mileages between towns, junctions & interchanges

Interstate Highways

U.S. Highways

State and Provincial Highways

Secondary State, Provincial and County Highways

Trans-Canada Highway, Canadian Autoroutes

Mexican Highways

Cities and Towns

Urbanized Areas

National Capitals; State Capitals

Cities and Towns; County Seats; Neighborhoods

Parks, Recreation Areas, Points of Interest

U.S. and Canadian National, State and Provincial Parks:

with camping facilities

without camping facilities

Points of Interest

Airports

Interchange Ramp Detail

Railroads

Corporate Limits

Larger Incorporated Areas

Buildings

City Parks

Golf Courses and Country Clubs

National Cemeteries

National Forests

National Parks

State Parks

Military Reservations, Points of Interest

Indian Reservations

Other Parks

Major North American Cities

MAJOR CITIES DATA Populations of cities, figure followed by (1990C), and their metropolitan areas (figure preceded by an asterisk) are the final census figures for the year cited. A population figure followed by (1990E) is an estimate for the year cited. Metropolitan areas have been defined by the Bureau of the Census and the U.S. Office of Management and Budget to include one or more major cities and their counties. A metropolitan area may also include additional counties that have strong economic and social ties to the central county or counties.

All information in the major cities guide has been checked for accuracy at the time of publication. Since changes do occur, the publisher cannot be responsible for any variations from the information printed.

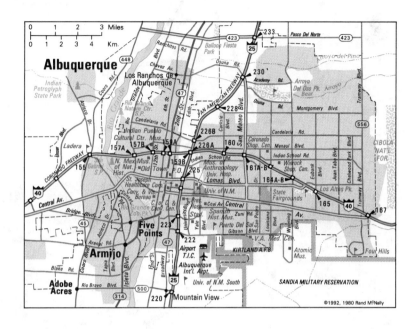

Albuquerque, New Mexico

Population: (*480,577) 384,736 (1990C)
Altitude: 4,958 feet
Average Temp.: Jan., 35°F.; July, 79°F.
Telephone Area Code: 505
Time: 247-1611 **Weather:** 821-1111
Time Zone: Mountain

AIRPORT TRANSPORTATION:

Eight miles to downtown Albuquerque. Taxicab, limousine, and city bus service.

SELECTED HOTELS:

Albuquerque Doubletree, 201 Marquette Ave. NW, 247-3344
Albuquerque Hilton Hotel, 1901 University Blvd. NE, 884-2500
Albuquerque Marriott Hotel, 2101 Louisiana Blvd. NE, 881-6800
AMFAC Hotel—Albuquerque Int'l Airport, 2910 Yale Blvd. SE, 843-7000
Best Western Airport Inn, 2400 Yale Blvd. SE, 242-7022
Best Western Winrock Inn, 18 Winrock Center, 883-5252
Holiday Inn—Midtown, 2020 Menaul Blvd. NE, 884-2511
Holiday Inn Pyramid—Journal Center, 5151 San Francisco Rd., 821-3333
Hyatt Regency, 330 Tijeras NW, 842-1234
La Posada, 125 2nd St. NW, 242-9090
Quality Hotel 4 Seasons, 2500 Carlisle NE, 888-3311
Ramada Hotel Classic, 6815 Menaul Blvd. NE, 881-0000
Ramada Inn East, 25 Hotel Circle NE, 296-5472
Sheraton—Old Town, 800 Rio Grande Blvd. NW, 843-6300

SELECTED RESTAURANTS:

The Cooperage, 7220 Lomas Blvd. NE, 255-1657
El Pinto, 10500 4th St. NW, 898-1771
Gardunòs, 10551 Montgomery NE, 298-5000
La Cascada, in the Albuquerque Doubletree Hotel, 247-3344
La Hacienda Restaurant & Cantina, 1306 Rio Grande Blvd. NW, 243-3709
Luna Mansion, Jct. US 85 & NM 6, 865-7333
Nicole's, in the Albuquerque Marriott Hotel, 881-6800
Prairie Star, on Jemez Canyon Rd., 867-3327
Rancher's Club, in the Albuquerque Hilton Hotel, 884-2500
Seagull Street, 5410 Academy NE, 821-0020

SELECTED ATTRACTIONS:

The Albuquerque Museum, 2000 Mountain Rd. NW, 242-4600
Cliffs Amusement Park, 4800 Osuna Rd. NE, 881-9373
Indian Pueblo Cultural Center, 2401 12th St. NW, 843-7270
Las Nutrias Vineyard & Winery, 4627 Corrales Rd., Corrales, 897-7863
National Atomic Museum, Wyoming Ave., Kirtland Air Force Base, 845-4636
New Mexico Museum of Natural History, 1801 Mountain Rd. NW, 841-8837
Petroglyph National Monument, 6900 Unser Blvd. NW, 873-6620
Rio Grande Nature Center State Park, 2901 Candelaria NW, 344-7240
Sandia Peak Aerial Tramway, #10 Tramway Loop NE, 298-8518
Turquoise Trail Scenic and Historic Area, North NM 14, the "Scenic Route" to Santa Fe, and NM 536 to the Sandia Crest, 281-5233

INFORMATION SOURCES:

Albuquerque Convention & Visitors Bureau
121 Tijeras NE
P.O. Box 26866
Albuquerque, New Mexico 87125-6866
(505) 243-3696; (800) 284-2282
Greater Albuquerque Chamber of Commerce
Albuquerque Convention Center
2nd & Marquette
Albuquerque, New Mexico 87102
(505) 764-3700

Atlanta, Georgia

Population: (*2,833,511) 394,017 (1990C)
Altitude: 1,050 feet
Average Temp.: Jan., 52°F.; July, 85°F.
Telephone Area Code: 404
Time: 936-8550 **Weather:** 767-1784
Time Zone: Eastern

AIRPORT TRANSPORTATION:

Eight miles to downtown Atlanta. Taxicab and limousine bus service.

SELECTED HOTELS:

Atlanta Hilton & Towers, 255 Courtland St. NE, 659-2000
Atlanta Marriott Marquis, 265 Peachtree Center Ave., 521-0000
Atlanta Marriott Perimeter Center, 246 Perimeter Center Pkwy., 394-6500
Best Western American Hotel, 160 Spring St., 688-8600
Colony Square Hotel, 188 14th St. NE, 892-6000
Holiday Inn—Airport North, 1380 Virginia Ave., 762-8411
Howard Johnson's North—Atlanta Airport, 1377 Virginia Ave., 762-5111
Hyatt Regency—Atlanta, 265 Peachtree St. NE, 577-1234

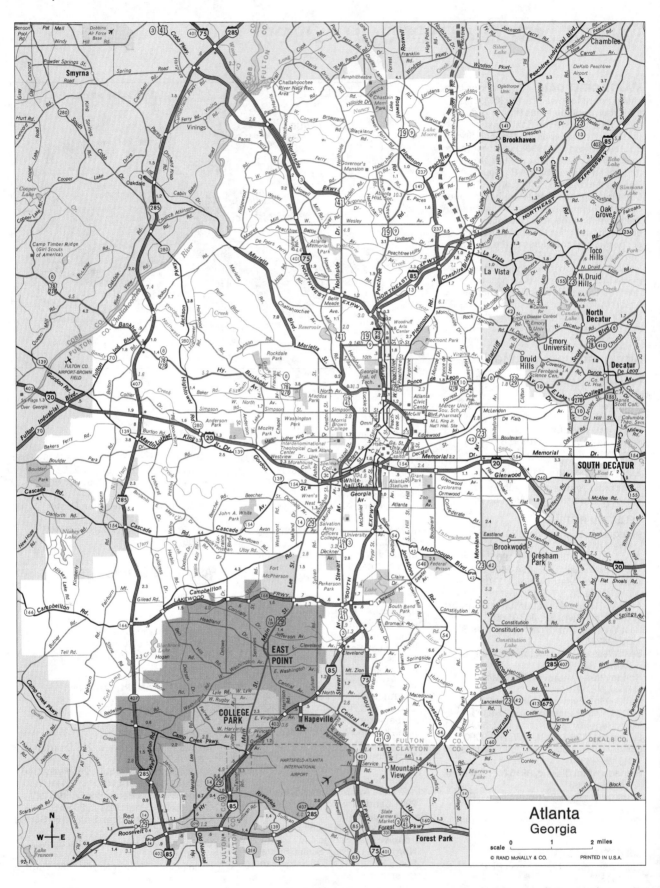

Atlanta
Georgia

scale 0 1 2 miles

© RAND McNALLY & CO.   PRINTED IN U.S.A.

92-1

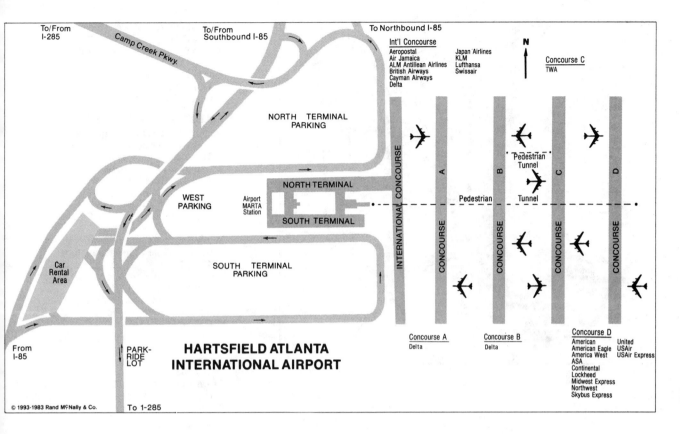

To/From
I-285

To/From
Southbound I-85

To Northbound I-85

Camp Creek Pkwy.

Int'l Concourse
Aeropostal
Air Jamaica
ALM Antillean Airlines
British Airways
Cayman Airways
Delta

Japan Airlines
KLM
Lufthansa
Swissair

N

Concourse C
TWA

NORTH TERMINAL
PARKING

INTERNATIONAL CONCOURSE

CONCOURSE A

CONCOURSE B

Pedestrian Tunnel

Pedestrian   Tunnel

CONCOURSE C

CONCOURSE D

WEST
PARKING

Airport
MARTA
Station

NORTH TERMINAL

SOUTH TERMINAL

Car
Rental
Area

SOUTH TERMINAL
PARKING

Concourse A
Delta

Concourse B
Delta

Concourse D
American   United
American Eagle USAir
America West   USAir Express
ASA
Continental
Lockheed
Midwest Express
Northwest
Skybus Express

From
I-85

PARK-
RIDE
LOT

**HARTSFIELD ATLANTA
INTERNATIONAL AIRPORT**

© 1993-1983 Rand McNally & Co.   To I-285

Omni Hotel at CNN Center, 100 CNN
   Center, 659-0000
Radisson Atlanta, 165 Courtland St.,
   659-6500
Ritz-Carlton Atlanta, 181 Peachtree St.
   NE, 659-0400
Stone Mountain Inn, U.S. 78, Stone
   Mountain Pk., 469-3311
The Westin Peachtree Plaza, 210
   Peachtree St. NW, 659-1400

SELECTED RESTAURANTS:
The Abbey, Piedmont between North &
   Ponce de Leon aves., 876-8831
Avanzare, in the Hyatt Regency—Atlanta,
   577-1234
Bugatti's, in the Omni Hotel at CNN
   Center, 659-0000
Cafe de la Paix, in the Atlanta Hilton &
   Towers, 659-2000
Coach And Six, 1776 Peachtree St. NW,
   872-6666
The Crab House, Piedmont at North Ave.
   (2nd Level—Rio), 872-0011
Daily's, 17 International Blvd., 681-3303
Dante's Down the Hatch, 3380 Peachtree
   Rd., 266-1600
La Grotta, 2637 Peachtree Rd. NE,
   231-1368
The Mansion, Piedmont at North Ave.,
   876-0727
Nikolai's Roof Restaurant, in the Atlanta
   Hilton & Towers, 659-2000
Pano's and Paul's, 1232 W. Paces Ferry
   Rd. NW, 261-3662
Terrace Garden Inn, 3405 Lenox Rd. NE,
   261-9250

SELECTED ATTRACTIONS:
Carter Presidential Center, 1 Copenhill,
   420-5110
Fernbank Museum of Natural History,
   767 Clifton Rd. NE, 378-0120
Fox Theatre, 660 Peachtree St. NE, (tours)
   522-4345
High Museum of Art, 1280 Peachtree St.
   NE, 892-3600
Martin Luther King Jr. Historic District,
   Auburn Ave. between Jackson &
   Randolph sts., (birth home) 331-3920,
   (Ebenezer Baptist Church) 688-7263,
   (grave & exhibit center) 524-1956
Stone Mountain Park, US 78, Stone
   Mountain, 498-5600
Underground Atlanta, Peachtree St. at
   Alabama St.
World of Coca-Cola Pavilion, 55 Martin
   Luther King Jr. Dr., 676-5151

INFORMATION SOURCES:
Atlanta Convention & Visitors Bureau
   Suite 2000, 233 Peachtree St. NE
   Atlanta, Georgia 30303
   (404) 521-6600
Atlanta Chamber of Commerce
   235 International Blvd. NW
   Atlanta, Georgia 30303
   (404) 880-9000

Austin, Texas

Population: (*781,572)
   465,622 (1990C)
Altitude: 501 feet
Average Temp.: Jan., 50°F.; July, 85°F.
Telephone Area Code: 512

Time: 973-3555 **Weather:** 476-7736
Time Zone: Central

AIRPORT TRANSPORTATION:
Seven-and-a-half miles to downtown
   Austin.
Taxicab, hotel shuttle, limousine and bus
   service.

SELECTED HOTELS:
Austin Crest Hotel, 111 E. 1st St.,
   478-9611
Austin Doubletree Hotel, 6505 I-35N,
   454-3737
Austin Marriott at the Capitol, 701 E. 11th
   St., 478-1111
Four Seasons Hotel—Austin, 98 San
   Jacinto Blvd., 478-4500
Guest Quarters, 303 W. 15th St., 478-7000
Hawthorn Suites—Northwest, 8888
   Tallwood Dr., 343-0008
Holiday Inn—Airport, 6911 N. I-35,
   459-4251
Howard Johnson's North Plaza, 7800
   Interregional Hwy. 35 N., 836-8520
Hyatt Regency—Austin, 208 Barton
   Springs, 477-1234
Red Lion Hotel, 6121 I-35N at US 290,
   323-5466
Wyndham Hotel Southpark, 4140
   Governor's Row, 448-2222

SELECTED RESTAURANTS:
County Line on the Hill, 6500 W. Bee
   Caves, 327-1742
County Line on the Lake, 5204 FM 2222,
   346-3664
Foothills of Austin, in the Hyatt
   Regency—Austin, 477-1234

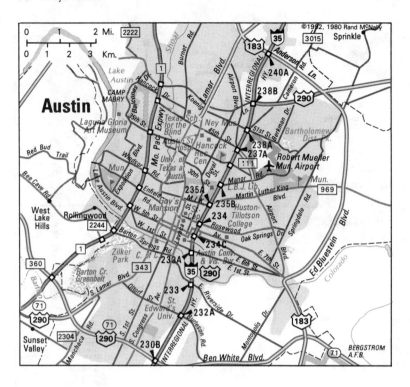

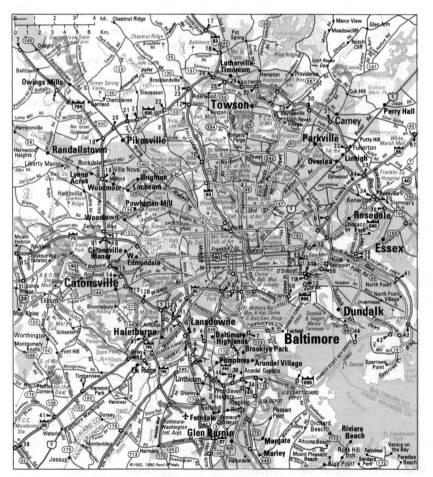

Green Pastures, 811 W. Live Oak,
444-4747
Matt's El Rancho, 2613 S. Lamar,
462-9333
Oasis Cantina Del Lago, 6550 Comanche
Trail, 266-2442
Threadgill's, 6416 N. Lamar, 451-5440

SELECTED ATTRACTIONS:

Barker Texas History Center, Sid
Richardson Hall, Unit 2, University of
Texas campus, 471-5961
Elisabet Ney Museum (sculpture), 304 E.
44th St., 458-2255
French Legation Museum, 802 San
Marcos at E. 7th St., 472-8180
Governor's Mansion, 1010 Colorado,
463-5518
Harry Ransom Center/Huntington Art
Gallery, West 21st & Guadalupe,
471-7324
Lyndon B. Johnson Presidential Library
and Museum, 2313 Red River,
University of Texas campus, 482-5279
Neill-Cochran House (ca 1855), 2310 San
Gabriel, 478-2335
O. Henry Home and Museum, 409 E. Fifth
St., 472-1903
State Capitol Building, 11th & Congress,
463-0063
Texas Memorial Museum, 2400 Trinity,
University of Texas campus, 471-1605
Treaty Oak, 503 Baylor

INFORMATION SOURCES:

Austin Convention & Visitors Bureau
900 Congress, Suite 300
Austin, Texas 78701
(512) 478-0098
Austin Visitors Center
Trask House
217 Red River
P. O. Box 1088
Austin, Texas 78769
Visitor Info. (512) 478-0098

Baltimore, Maryland

Population: (*2,382,172)
736,014 (1990C)
Altitude: Sea level to 32 feet
Average Temp.: Jan., 37°F.; July, 79°F.
Telephone Area Code: 410
Time: 844-2525   **Weather:** 936-1212
Time Zone: Eastern

AIRPORT TRANSPORTATION:

Ten miles to downtown Baltimore.
Taxicab and limousine bus service.

SELECTED HOTELS:

Baltimore Ramada Hotel, 1701 Belmont
Ave. at Security Blvd., 265-1100
Belvedere Hotel, 1 E. Chase St., 332-1000
Brookshire Hotel, 120 E. Lombard St.,
625-1300
Cross Keys Inn, 5100 Falls Rd., 532-6900
Harbor Court Hotel, 550 Light St.,
234-0550
Holiday Inn at the Harbor, 301 W.
Lombard St., 685-3500
Holiday Inn—BWI Airport, 890 Elkridge
Landing Rd., 859-8400
Hyatt Regency Baltimore, 300 Light St.,
528-1234

Marriott Inner Harbor, Pratt & Eutaw sts.,
962-0202
Marriott's Hunt Valley Inn, 245 Shawan
Rd., Hunt Valley, 785-7000
Omni Inner Harbor Hotel, 101 W. Fayette
St., 752-1100
Peabody Court Hotel, 612 Cathedral St.,
727-7101
Pikesville Hilton Inn, 1726 Reisterstown
Rd. at the Beltway, Pikesville, 653-1100
Sheraton International Hotel, 7032 Elm
Rd., at Baltimore/Washington Internat'l
Airport, 859-3300
Stouffer Harborplace Hotel, 202 E. Pratt
St., 547-1200

SELECTED RESTAURANTS:
Bamboo House, Harborplace—Pratt St.
Pavilion, 625-1191
Chiapparelli's, 237 S. High St., 837-0309
The Conservatory, in the Peabody Court
Hotel, 727-7101
Crossroads, in Cross Keys Inn Hotel,
532-6900
Hampton's, in the Harbor Court Hotel,
234-0550
Haussner's Restaurant, 3242 Eastern
Ave., 327-8365
Obrycki's Crab House, 1727 E. Pratt St.,
732-6399
The Prime Rib, 1101 N. Calvert St.,
539-1804
Tio Pepe Restaurant, 10 E. Franklin St.,
539-4675

SELECTED ATTRACTIONS:
B&O Railroad Museum, 901 W. Pratt St.,
752-2490
Baltimore Museum of Art, Art Museum
Dr., 396-7101
Baltimore Zoo, Druid Hill Park, 366-5466
Fort McHenry National Monument and
Historic Shrine, Foot of E. Fort Ave.,
962-4290
Harbor Cruises, Ltd., 301 Light St.,
727-3113
Harborplace/The Gallery at Harborplace,
Light & Pratt sts. and Calvert & Pratt
sts., 332-4191
Lexington Market, Lexington & Eutaw
sts., 685-6169
Maryland Science Center, 601 Light St.,
685-5225
National Aquarium in Baltimore, Pier 3,
501 E. Pratt St., 576-3800
Top of the World Trade Center, 401 E.
Pratt St., 837-4515

INFORMATION SOURCES:
Baltimore Area Convention and Visitors
Association
1 E. Pratt St., Plaza Level
Baltimore, Maryland 21202
(410) 659-7300
Visitor Info. (410) 837-4636;
(800) 282-6632 (except Baltimore)
Baltimore Office of Promotion
200 W. Lombard St.
Baltimore, Maryland 21201
(410) 752-8631
Greater Baltimore Committee
Legg Mason Tower
111 S. Calvert St., Suite 1500
Baltimore, Maryland 21202
(410) 727-2820

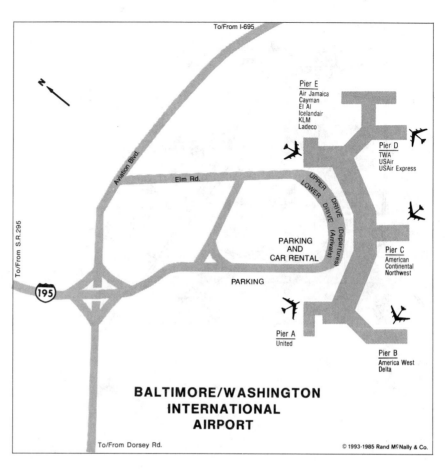

**BALTIMORE/WASHINGTON
INTERNATIONAL
AIRPORT**

© 1993-1985 Rand McNally & Co.

Baton Rouge, Louisiana

Population: (*528,264)
219,531 (1990C)
Altitude: 58 feet
Average Temp.: Jan., 51°F.; July, 82°F.
Telephone Area Code: 504
Time: 387-5411 **Weather:** 355-4823
Time Zone: Central

AIRPORT TRANSPORTATION:
Seven miles to downtown Baton Rouge.
Taxicab, hotel van, and limousine
service.

SELECTED HOTELS:
Baton Rouge Hilton, 5500 Hilton Ave.,
924-5000
Bellemont Hotel, 7370 Airline Hwy.,
357-8612
Courtyard By Marriott, 2421 S. Acadian
Thruway, 924-6400
Holiday Inn—East, I-10 & Siegen Ln.,
293-6880
Holiday Inn—South, 9940 Airline Hwy.,
924-7021
Quality Inn—Baton Rouge, 10920 Mead
Rd., 293-9370
Ramada Hotel, 1480 Nicholson Dr.,
387-1111
Sheraton Baton Rouge, 4728 Constitution
Ave., 925-2244

SELECTED RESTAURANTS:
Chalet Brandt, 7655 Old Hammond Hwy.,
927-6040
The Chinese Restaurant, 1710 Nicholson
Dr., 387-9443
Don's Seafood & Steakhouse, 6823
Airline Hwy., 357-0601
Giamanco's, 4624 Government St.,
928-5045
Juban's, 3739 Perkins Rd., 346-8422
Maison LaCour, 11025 N. Harrell's Ferry
Rd., 275-3755
Mike Anderson's Seafood, 1031 W. Lee
Dr., 766-3728
Mulate's Cajun Restaurant, 8322
Bluebonnet Rd., 767-4794
Ruth's Chris Steak House, 4836
Constitution, 925-0163

SELECTED ATTRACTIONS:
Louisiana Arts & Science
Center/Riverside Museum, North Blvd.
& River Rd., 344-9463
Louisiana State Capitol, State Capitol Dr.,
342-7317
Louisiana State University (museums,
Indian mounds), south of downtown
between Highland Rd. and Nicholson
Dr., 388-3202
LSU Rural Life Museum (19th-century
buildings), Essen Ln. at I-10, 765-2437
McGee's Atchafalaya Basin Tours, the
Atchafalaya Swamps in Henderson,
(318) 228-8519

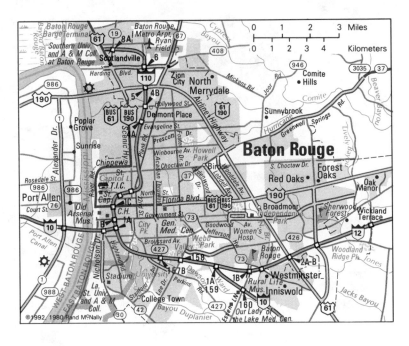

Magnolia Mound Plantation, 2161
  Nicholson Dr., 343-4955
Nottoway Plantation, LA Hwy. 1, White
  Castle, 346-8263
Oak Alley Plantation, LA Hwy. 18,
  Vacherie, 265-2151
Old State Capitol, North Blvd. at River
  Rd., 342-8211
U.S.S. *Kidd*/Nautical Historic Center,
  Government St. at River Rd., 342-1942

INFORMATION SOURCES:

Baton Rouge Area Convention & Visitors
Bureau
  730 North Boulevard
  Baton Rouge, Louisiana 70802
  (504) 383-1825
The Greater Baton Rouge Chamber of
Commerce
  564 Laurel St.
  P.O. Box 3217
  Baton Rouge, Louisiana 70821
  (504) 381-7125

Birmingham, Alabama

Population: (*907,810)
  265,968 (1990C)
Altitude: 601 feet
Average Temp.: Jan., 46°F.; July, 82°F.
Telephone Area Code: 205
Time: 979-8463 **Weather:** 945-7000
Time Zone: Central

AIRPORT TRANSPORTATION:

Five miles to downtown Birmingham.
Taxicab, bus and limousine service.

SELECTED HOTELS:

Holiday Inn—Airport, 5000 10th Ave. N.,
  591-6900
Holiday Inn—East, 7941 Crestwood Blvd.,
  956-8211
Holiday Inn—Galleria Area, 1548
  Montgomery Hwy., 822-4350
Holiday Inn Homewood, 260 Oxmoor Rd.,
  942-2041
Mountain Brook Inn, 2800 US 280,
  870-3100
Radisson Birmingham, 808 S. 20th St. at
  University Blvd., 933-9000
Ramada Inn—Airport, 5216 Airport Hwy.,
  591-7900
Ramada Inn Central, 300 N. 10th St.,
  328-8560
The Tutwiler, Park Place at 21st St. N.,
  322-2100
Wynfrey Hotel, 1000 Riverchase Galleria,
  987-1600

SELECTED RESTAURANTS:

Bombay Cafe, 2839 7th Ave. S., 322-1930
Christian's Classic Cuisine, 2300
  Woodcrest Pl., 871-3222
GG in the Park, 3625 8th Ave. S., 254-3506
Highlands: A Bar & Grill, 2011 11th Ave.
  S., 939-1400
John's Restaurant, 112 N. 21st St.,
  322-6014
Michael's Sirloin Room, 431 S. 20th St.,
  322-0419
Winston's, in the Wynfrey Hotel,
  987-1600

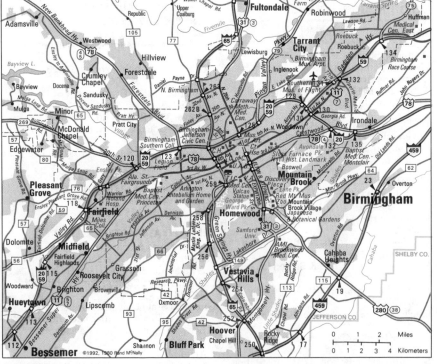

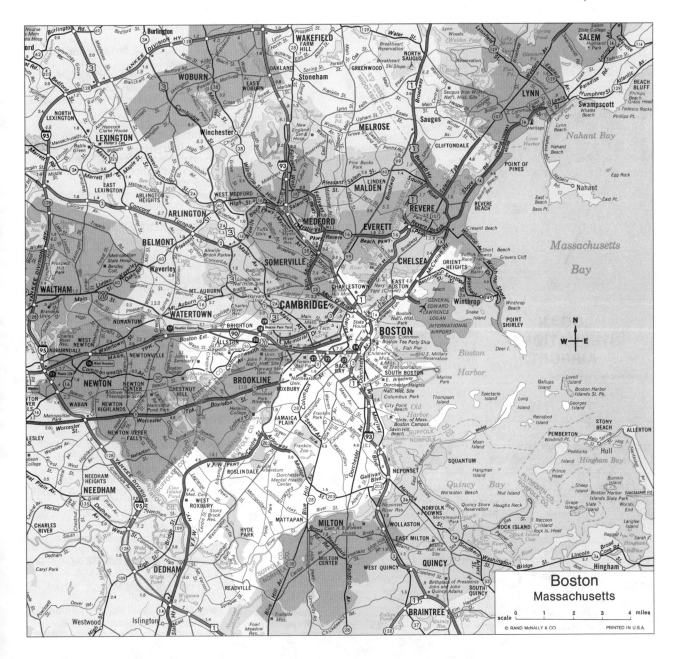

Boston
Massachusetts

scale 0 1 2 3 4 miles

© RAND McNALLY & CO. PRINTED IN U.S.A.

SELECTED ATTRACTIONS:

Alabama Sports Hall of Fame Museum, 22nd St. & Civic Center Blvd., 323-6665

Arlington Antebellum Home and Gardens, 331 Cotton Ave. SW, 780-5656

Birmingham Museum of Art, 2000 8th Ave. N., 254-2565

Birmingham Zoo, 2630 Cahaba Rd., 879-0408

Botanical and Japanese Gardens, 2612 Lane Park Rd., 879-1227

Discovery Place (children's museum), 1320 22nd St. S., 939-1176

Five Points South (historic district), 11th Ave. S. & 20th St.

Red Mountain Museum (earth's history), 2230 Arlington Crescent, 933-4153

Riverchase Galleria, U.S. 31 South at I-459, 985-3039

Sloss Furnaces (iron-making), beside First Ave. —north viaduct on 32nd St., 324-1911

Vulcan Park (panoramic view from world's largest cast iron statue), Valley Ave. at Highway 31 S., 328-6198

INFORMATION SOURCES:

Greater Birmingham Convention and Visitors Bureau
2200 Ninth Ave. N.
Birmingham, Alabama 35203
(205) 252-9825

Birmingham Area Chamber of Commerce
2027 First Ave. N., Suite 1200
North Birmingham, Alabama 35203
(205) 323-5461

Boston, Massachusetts

Population: (*2,870,669)
574,283 (1990C)
Altitude: Sea level to 330 feet
Average Temp.: Jan., 29°F.; July, 72°F.
Telephone Area Code: 617
Time: 637-1234 **Weather:** 936-1234
Time Zone: Eastern

AIRPORT TRANSPORTATION:

Three miles to downtown Boston.
Taxicab and limousine bus service.

SELECTED HOTELS:

Boston Harbor Hotel, 70 Rowes Wharf, 439-7000

Boston Marriott, Copley Place, 110 Huntington Ave., 236-5800

The Colonnade, 120 Huntington Ave., 424-7000

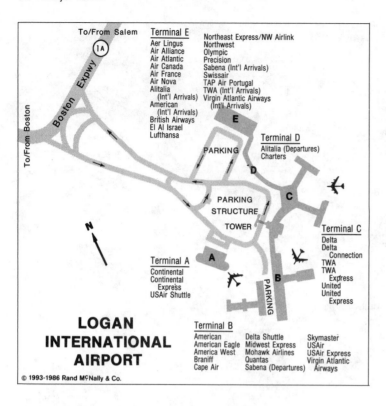

To/From Salem

Terminal E
Aer Lingus
Air Alliance
Air Atlantic
Air Canada
Air France
Air Nova
Alitalia
 (Int'l Arrivals)
American
 (Int'l Arrivals)
British Airways
El Al Israel
Lufthansa

Northeast Express/NW Airlink
Northwest
Olympic
Precision
Sabena (Int'l Arrivals)
Swissair
TAP Air Portugal
TWA (Int'l Arrivals)
Virgin Atlantic Airways
 (Int'l Arrivals)

Terminal D
Alitalia (Departures)
Charters

Terminal C
Delta
Delta
 Connection
TWA
TWA
 Express
United
United
 Express

Terminal A
Continental
Continental
 Express
USAir Shuttle

Terminal B
American
American Eagle
America West
Braniff
Cape Air

Delta Shuttle
Midwest Express
Mohawk Airlines
Quantas
Sabena (Departures)

Skymaster
USAir
USAir Express
Virgin Atlantic
 Airways

LOGAN INTERNATIONAL AIRPORT
© 1993-1986 Rand McNally & Co.

Copley Plaza Hotel, 138 St. James Ave., 267-5300
The Four Seasons, 200 Boylston St., 338-4400
Hotel Meridien, 250 Franklin St., 451-1900
Logan Airport Hilton, 75 Service Rd., Logan International Airport, 569-9300
Omni Parker House, 60 School St., 227-8600
The Ritz-Carlton, Boston, 15 Arlington St., 536-5700
Sheraton Boston Hotel & Towers, Prudential Center, 39 Dalton St., 236-2000
Westin Hotel Copley Place, 10 Huntington Ave., 262-9600

SELECTED RESTAURANTS:
Anthony's Pier 4, 140 Northern Ave., 423-6363
The Cafe Budapest, 90 Exeter St., 266-1979
Caffe Lampara, 916 Commonwealth Ave., 566-0300
Copley's Restaurant, in the Copley Plaza Hotel, 267-5300
The Dining Room, in The Ritz-Carlton, Boston, 536-5700
Felicia's, 145A Richmond St., up one flight, 523-9885
Genji, 327 Newbury St., 267-5656
Hampshire House, 84 Beacon St., 227-9600
Jimmy's Harborside Restaurant, 242 Northern Ave., 423-1000
Julien, in the Hotel Meridien, 451-1900
Legal Seafoods, in the Boston Park Plaza Hotel, 35 Columbus Ave., 426-4444
Locke—Ober Cafe, 3 Winter Pl., 542-1340
Maison Robert, Old City Hall, 45 School St., 227-3370

SELECTED ATTRACTIONS:
Boston Tea Party Ship & Museum, Congress St. Bridge, 338-1773
Cheers!/Bull & Finch Pub, 84 Beacon St., 227-9605
Children's Museum, Museum Wharf, 300 Congress St., 426-8855
Faneuil Hall/Faneuil Hall Marketplace, Congress & North sts., 523-1300
The Freedom Trail (self-guided walking tour), 242-5642
Harvard University, Cambridge, 495-1000
John F. Kennedy Library & Museum, Columbia Point, Dorchester, 929-4523
Museum of Fine Arts, 465 Huntington Ave., 267-9300
Museum of Science, 1 Science Park, 723-2500
New England Aquarium, Central Wharf, 973-5200

INFORMATION SOURCES:
Greater Boston Convention & Visitors Bureau, Inc.
 Prudential Tower, Suite 400, Box 490
 Boston, Massachusetts 02199
 (617) 536-4100
Greater Boston Chamber of Commerce
 600 Atlantic Ave.
 Boston, Massachusetts 02210-2200
 (617) 227-4500

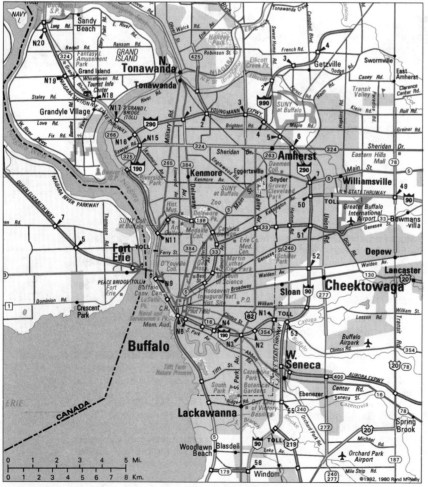

Buffalo, New York

Population: (*968,532)
   328,123 (1990C)
Altitude: 600 feet
Average Temp.: Jan., 26°F.; July, 71°F.
Telephone Area Code: 716
Time: 844-1717 **Weather:** 844-4444
Time Zone: Eastern

AIRPORT TRANSPORTATION:

Nine miles to downtown Buffalo.
Taxicab and limousine bus service.

SELECTED HOTELS:

Best Western Inn—Downtown, 510
   Delaware Ave., 886-8333
Buffalo Hilton, 120 Church St., 845-5100
Buffalo Marriott Inn, 1340 Millersport
   Hwy., Amherst, 689-6900
Holiday Inn—Buffalo Airport, 4600
   Genesee St., Cheektowaga, 634-6969
Holiday Inn—Buffalo Downtown, 620
   Delaware Ave., 886-2121
Holiday Inn—Gateway, 601 Dingens St.,
   896-2900
Howard Johnson's Motor Lodge, 6700
   Transit Rd., Williamsville, 634-7500
Hyatt Regency, Pearl & W. Huron,
   856-1234
Quality Inn, 4217 Genesee St.,
   Cheektowaga, 633-5500
Radisson Hotel & Suites, 4243 Genesee
   St., 634-2300
Ramada Inn—Buffalo Airport, 6643
   Transit Rd., 634-2700
Sheraton Buffalo, 2040 Walden Ave.,
   Cheektowaga, 681-2400
Williamsville Inn, 5447 Main St.,
   Williamsville, 634-1111

SELECTED RESTAURANTS:

Asa Ransom House, 10529 Main St.,
   Clarence, 759-2315
Daffodil's Restaurant, 930 Maple Rd.,
   Williamsville, 688-5413
E.B. Greens, in the Hyatt Regency,
   856-1234
Grill Ninety-One, 91 Niagara St., 856-8373
Justine's, in the Buffalo Hilton, 845-5100
Lord Chumley's, 481 Delaware Ave.,
   886-2220
Old Red Mill Inn, 8326 Main St.,
   Williamsville, 633-7878
Park Lane at the Circle, Delaware Ave. at
   Gates Circle, 883-3344

SELECTED ATTRACTIONS:

Albright-Knox Art Gallery, 1285 Elmwood
   Ave., 882-8700
Buffalo and Erie County Historical
   Society, 25 Nottingham Ct., 873-9644
Buffalo Museum of Science, 1020
   Humboldt Pkwy., 896-5200
Buffalo Naval & Servicemen's Park, 1
   Naval Park Cove, 847-1773
Buffalo Zoo, across from Delaware Park,
   837-3900
Historic Theatre District, 856-3150
Miss Buffalo-Niagara Clipper, 856-6696
The Original American Kazoo Company,
   8703 South Main St., Eden, 992-3960

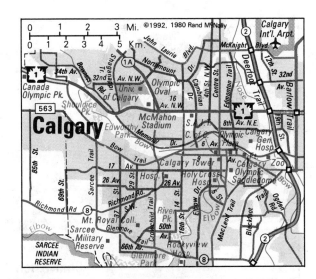

INFORMATION SOURCES:

Greater Buffalo Convention and Visitors
Bureau
   107 Delaware Ave.
   Buffalo, New York 14202
   (716) 852-0511; (800) 283-3256
Greater Buffalo Chamber of Commerce
   107 Delaware Ave.
   Buffalo, New York 14202
   (716) 852-7100

Calgary, Alberta, Canada

Population: (*671,326)
   636,104 (1986C)
Altitude: 3,439 feet
Average Temp.: Jan., 13°F.; July, 62°F.
Telephone Area Code: 403
Time: 263-3333 **Weather:** 275-3300
Time Zone: Mountain

AIRPORT TRANSPORTATION:

Ten miles to downtown Calgary.
Taxicab, airporter shuttle, and limousine
   service.

SELECTED HOTELS:

Best Western Hospitality Inn, 135
   Southland Dr. SE, 278-5050
Blackfoot Inn, 5940 Blackfoot Trail SE,
   252-2253
Chateau Airport Hotel, 2001 Airport Rd.
   NE, 291-2600
Delta Bow Valley, 209 4th Ave. SE,
   266-1980
The Palliser, 9th Ave. & 1st St. SW,
   262-1234
Port O'Call Inn, 1935 McKnight Blvd. NE,
   291-4600
Prince Royal Inn, 618 5th Ave. SW,
   263-0520
Ramada Downtown, 708 8th Ave. SW,
   263-7600
The Sandman, 888 7th Ave. SW, 237-8626
Sheraton Cavalier, 2620 32nd Ave. NE,
   291-0107
Skyline Plaza, 110 9th Ave. SE, 266-7331
The Westin Hotel, 320 4th Ave. SW,
   266-1611

SELECTED RESTAURANTS:

Atrium Terrace, in the Chateau Airport
   Hotel, 291-2600
Caesar's, 512 4th Ave. SW, 264-1222
Hy's Steak House, 316 4th Ave. SW,
   263-2222
Japanese Village, 302 4th Ave. SW,
   262-2738
La Caille on the Bow, 805 1st Ave. SW,
   262-5554
La Chaumiere, 121 17th Ave. SE,
   228-5690
La Dolce Vita, 916 1st Ave. NE, 263-3445
Owl's Nest Dining Room, in the Westin
   Hotel, 266-1611
Panorama Room, atop the Calgary
   Tower, 101 9th Ave. SW, 266-7171
Rimrock Room, in the Palliser Hotel,
   262-1234

SELECTED ATTRACTIONS:

Alberta Science Centre/Centennial
   Planetarium, 701 11th St. SW, 221-3700
Calaway Park, 10 km. west on the
   Trans-Canada Hwy., 240-3822
Calgary Tower, 9th Ave. and Centre St.,
   266-7171
Calgary Zoo, Botanical Gardens &
   Prehistoric Park, Memorial Dr. west of
   Deerfoot Trail, 232-9372
Canada Olympic Park, on the western city
   limits along the Trans-Canada Hwy.,
   286-2632
Fort Calgary, 750 9th Ave. SE, 290-1875
Glenbow Museum, 130 9th Ave. SE,
   264-8300
Heritage Park (turn-of-the-century
   village), Heritage Dr. & 14th St. SW,
   259-1900
Inglewood Bird Sanctuary, south of 9th
   Ave. & Sanctuary Rd. SE, 269-6688
Museum of Movie Art, #9 3600 21st St.
   NE, 250-7588

INFORMATION SOURCES:

Calgary Convention & Visitors Bureau
   237 8th Ave. SE
   Calgary, Alberta, Canada T2G 0K8
   (403) 263-8510

The Calgary Chamber of Commerce
517 Centre St. S.
Calgary, Alberta, Canada T2G 2C4
(403) 263-7435

Charleston, South Carolina

Population: (*506,875)
80,414 (1990C)
Altitude: 118 feet
Average Temp.: Jan., 49°F.; July, 80°F.
Telephone Area Code: 803
Time: 572-8463 **Weather:** 744-3207
Time Zone: Eastern

AIRPORT TRANSPORTATION:

Twelve miles to downtown Charleston.
Taxicab, airport shuttle, hotel van, and
limousine service.

SELECTED HOTELS:

Charleston Marriott, I-26 at Montague
Ave., 747-1900
Comfort Inn Riverview, 144 Bee St.,
577-2224
Days Inn, 155 Meeting St., 722-8411
Hampton Inn—Riverview Hotel, 11
Ashley Pointe Dr., 556-5200
Holiday Inn International Airport, I-26 &
W. Aviation, 744-1621
Holiday Inn—Mt. Pleasant, 250 Johnnie
Dodds Blvd., Mt. Pleasant, 884-6000
Holiday Inn—Riverview, 301 Savannah
Hwy., 556-7100
Mills House Hotel, 115 Meeting St.,
577-2400
Northwoods Atrium Inn—Best Western,
7401 Northwoods Blvd., 572-2200
Omni Hotel at Charleston Place, 130
Market St., 722-4900
Quality Inn Heart of Charleston, 125
Calhoun St., 722-3391
Ramada Inn Airport, I-26 & W. Montague,
744-8281
Sheraton Airport Inn, I-26 & E. Aviation
Ave., 744-2501

Sheraton Charleston, 170 Lockwood Dr.,
723-3000

SELECTED RESTAURANTS:

Barbadoes Room, in the Mill House
Hotel, 577-2400
Carolina's, 10 Exchange St., 724-3800
The Colony House, 35 Prioleau St.,
723-3424
French Quarter, in the Lodge Alley Inn,
195 E. Bay St., 722-1611
Henry's, 54 North Market St., 723-4363
La Grill, 32 Market St., 723-3614
Marianne, 235 Meeting St., 722-7196
The Palmetto Cafe, in the Omni Hotel at
Charleston Place, 722-4900
Shem Creek Bar & Grill, 508 Mill St., Mt.
Pleasant, 884-8102

SELECTED ATTRACTIONS:

Audubon Swamp Garden, 10 mi. NW on
SC Hwy. 61, enter at Magnolia
Plantation, 571-1266
The Charleston Museum, 360 Meeting
St., 722-2996
Charles Towne Landing 1670, 1500 Old
Town Rd., 556-4450
Drayton Hall, 3380 Ashley River Rd.,
766-0188
Fort Moultrie, West Middle St. on
Sullivan's Island, 883-3123
Fort Sumter tours, City Marina, 722-1691
Gibbes Museum of Art, 135 Meeting St.,
722-2706
Heyward-Washington House, 87 Church
St., 722-0354
Magnolia Plantation & Gardens, 10 mi.
NW on SC Hwy. 61, 571-1266
Patriots Point Naval & Maritime Museum,
40 Patriots Point Rd., Mt. Pleasant,
884-2727

INFORMATION SOURCES:

Charleston Trident Convention & Visitors
Bureau
81 Mary St.
P.O. Box 975

Charleston, South Carolina 29402
(803) 853-8000
Visitor Reception & Transportation
Center
375 Meeting St.
Charleston, South Carolina 29403
(803) 853-8000
Charleston Trident Chamber of
Commerce
81 Mary St.
Charleston, South Carolina 29403
(803) 577-2510

Charlotte, North Carolina

Population: (*1,162,093)
395,934 (1990C)
Altitude: 700 feet
Average Temp.: Jan., 52°F.; July, 88°F.
Telephone Area Code: 704
Time: 375-6711 **Weather:** 359-8466
Time Zone: Eastern

AIRPORT TRANSPORTATION:

Ten miles to downtown Charlotte.
Taxicab and hotel shuttle service.

SELECTED HOTELS:

Adam's Mark Charlotte, 555 S. McDowell
St., 372-4100
Charlotte Marriott City Center, 100 W.
Trade, 333-9000
Embassy Suites, 4800 S. Tryon, 527-8400
Holiday Inn—Sugar Creek, at I-85 Sugar
Creek exit, 596-9390
Howard Johnson's—South, 118 E.
Woodlawn Rd., 525-6220
Omni Charlotte, 222 E. 3rd., 377-6664
Quality Inn Airport, I-85 at Little Rock Rd.,
394-4111
Radisson Plaza, 2 Nations Bank Plaza at
Trade & College sts., 377-0400
Ramada Inn Independence, 3501 E.
Independence Blvd., 537-1010
The Registry Hotel, 321 W. Woodlawn
Rd., 525-4441

SELECTED RESTAURANTS:

Azalea's, in the Radisson Plaza, 377-0400
Epicurean, 1324 East Blvd., 377-4529
Hereford Barn Steak House, 4320 N. I-85
Service Rd., 596-0854
The Lamplighter, 1065 E. Morehead,
372-5343
Ranch House, 5614 Wilkinson Blvd.,
399-5411
The Registry Café, in The Registry Hotel,
525-4441
Slug's 30th Edition, 2 First Union Tower,
30th Floor, 372-7778

SELECTED ATTRACTIONS:

Carowinds (theme park), I-77 &
Carowinds Blvd. (exit 90), 588-2606
Charlotte Motor Speedway, Concord,
455-3200
Discovery Place (science museum,
Omnimax Theatre & Planetarium), 301
N. Tryon St., 372-6261
Mint Museum of Art, 2730 Randolph Rd.,
337-2000
Reed Gold Mine State Historic Site, 20
mi. east of Charlotte off Albemarle Rd.
(NC Hwy. 24/27), 786-8337

INFORMATION SOURCES:

Charlotte Convention & Visitors Bureau
  229 North Church St.
  Charlotte, North Carolina 28202
  (704) 334-2282
Charlotte Chamber of Commerce
  129 W. Trade St.
  P. O. Box 32785
  Charlotte, North Carolina 28232
  (704) 377-6911

Chicago, Illinois

Population: (*6,069,974)
  2,783,726 (1990C)
Altitude: 596 feet
Average Temp.: Jan., 27°F.; July, 75°F.
Telephone Area Code: 312
Time: 976-6000 **Weather:** 976-1212
Time Zone: Central

AIRPORT TRANSPORTATION:

Nineteen miles from O'Hare to
  downtown Chicago; 10 miles from
  Midway to downtown Chicago.
Taxicab and limousine bus service from
  both airports; also bus/rapid transit
  from O'Hare.

SELECTED HOTELS:

Ambassador West Hotel, 1300 N. State
  Pkwy., 787-7900
Barclay Chicago, 166 E. Superior,
  787-6000
Chicago Marriott Hotel, 540 N. Michigan
  Ave., 836-0100
Days Inn, 644 N. Lake Shore Dr., 943-9200
The Drake, 140 E. Walton Place, 787-2200
Fairmont Hotel, 200 N. Columbus Dr. at
  Illinois Center, 565-8000
Holiday Inn, 350 N. Orleans St., 836-5000
Hotel Nikko, 320 N. Dearborn, 744-1900
Hyatt Regency Chicago, 151 E. Wacker
  Dr., 565-1234
Hyatt Regency O'Hare, 9300 W. Bryn
  Mawr Ave., Rosemont, (708) 696-1234
Marriott O'Hare, 8535 W. Higgins Rd.,
  693-4444
Mayfair Regent Hotel, 181 E. Lake Shore
  Dr., 787-8500
Omni Ambassador East Hotel, 1301 N.
  State Pkwy., 787-7200
Palmer House Hilton, 17 E. Monroe St.,
  726-7500
Park Hyatt, Chicago Ave. & Rush St.,
  280-2222
The Ritz-Carlton, Chicago, 160 E. Pearson
  St., 266-1000
Swissotel Chicago, 323 E. Wacker,
  565-0565
Tremont Hotel, 100 E. Chestnut St.,
  751-1900
The Westin Hotel, 909 N. Michigan Ave.,
  943-7200

SELECTED RESTAURANTS:

Biggs Restaurant, 1150 N. Dearborn
  Pkwy., 787-0900
Cape Cod Room, in the Drake Hotel,
  787-2200
Cricket's, in the Tremont Hotel, 751-1900
Gordon Restaurant, 500 N. Clark,
  467-9780
House of Hunan, 535 N. Michigan Ave.,
  329-9494
La Tour, in the Park Hyatt, 280-2222

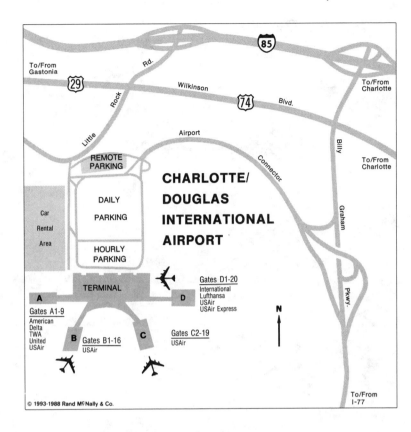

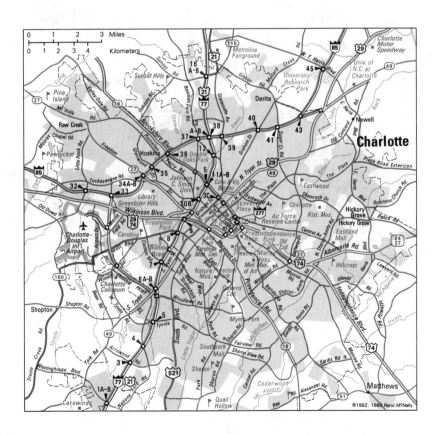

Chicago
Illinois

scale 0 1 2 3 4 5 miles

© RAND McNALLY & CO. PRINTED IN U.S.A.

N W E S

Lake Michigan

HIGHLAND PARK

Lake Zurich
Prairie View
Half Day
Bannockburn
Deerfield
Northbrook
Glencoe
Wheeling
Buffalo Gr.
Dundee
Northfield
Winnetka
Glenview
WILMETTE
Kenilworth
Palatine
ARLINGTON HTS.
Prospect Hts.
MT. PROSPECT
GLENVIEW
MORTON GR.
SKOKIE
EVANSTON
Rolling Meadows
DES PLAINES
NILES
Lincolnwood
HOFFMAN ESTATES
SCHAUMBURG
PARK RIDGE
ELK GROVE VIL.
Rosemont
CHICAGO O'HARE INTERNATIONAL AIRPORT
Itasca
Wood Dale
Bensenville
Schiller Park
Norridge
Harwood Hts.
Roselle
Medinah
Bloomingdale
Addison
Franklin Park
River Gr.
Elmwood Park
Lincoln Park Zoo
De Paul University
Carol Stream
Glen Ellyn Countryside
Glendale Hts.
Villa Park
MELROSE PARK
Stone Park
OAK PARK
CHICAGO
Glen Ellyn
LOMBARD
ELMHURST
Berkeley
River Forest
Chicago Historical Society
LAKE SHORE PARK
Northwestern University (Chicago Campus)
WHEATON
Hillside
Bellwood
MAYWOOD
Forest Park
BERWYN
CICERO
Grant Park
Buckingham Fountain
Shedd Aquarium
Adler Planetarium
Westchester
N. Riverside
Riverside
Field Museum
Merrill C. Meigs Airport
Mc Cormick Place
Oak Brook
LaGrange Park
Lyons
Stickney
Lisle
DOWNERS GR.
Clarendon Hills
Hinsdale
Western Sprs.
Brookfield
Summit
NAPERVILLE
Darien
Woodridge
Burr Ridge
Justice
Bridgeview
BURBANK
Hometown
Museum of Science and Industry
JACKSON PARK
BOLINGBROOK
Lemont
Willow Sprs.
Hickory Hills
OAK LAWN
Evergreen Park
Chicago Ridge
Worth
Palos Hills
Alsip
Calumet Park
Blue Island
Romeoville
Orland Park
Palos Park
Palos Hts.
Crestwood
Robbins
Posen
Riverdale
Dolton
Burnham
Midlothian
Oak Forest
HARVEY
Phoenix
SOUTH HOLLAND
CALUMET CITY
Tinley Creek Woods
Markham

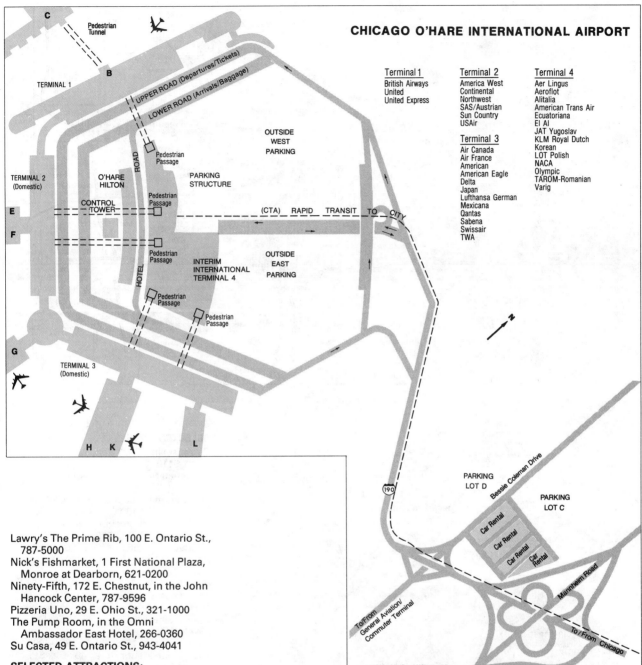

CHICAGO O'HARE INTERNATIONAL AIRPORT

Terminal 1
British Airways
United
United Express

Terminal 2
America West
Continental
Northwest
SAS/Austrian
Sun Country
USAir

Terminal 3
Air Canada
Air France
American
American Eagle
Delta
Japan
Lufthansa German
Mexicana
Qantas
Sabena
Swissair
TWA

Terminal 4
Aer Lingus
Aeroflot
Alitalia
American Trans Air
Ecuatoriana
El Al
JAT Yugoslav
KLM Royal Dutch
Korean
LOT Polish
NACA
Olympic
TAROM-Romanian
Varig

TERMINAL 1

TERMINAL 2 (Domestic)

TERMINAL 3 (Domestic)

Pedestrian Tunnel

UPPER ROAD (Departures/Tickets)

LOWER ROAD (Arrivals/Baggage)

O'HARE HILTON

CONTROL TOWER

ROAD

HOTEL

Pedestrian Passage

OUTSIDE WEST PARKING

PARKING STRUCTURE

(CTA) RAPID TRANSIT TO CITY

INTERIM INTERNATIONAL TERMINAL 4

OUTSIDE EAST PARKING

PARKING LOT D

PARKING LOT C

Bessie Coleman Drive

Car Rental

Mannheim Road

To/From General Aviation/ Commuter Terminal

To / From Chicago

© 1993-1983 Rand McNally & Co.

Lawry's The Prime Rib, 100 E. Ontario St.,
787-5000
Nick's Fishmarket, 1 First National Plaza,
Monroe at Dearborn, 621-0200
Ninety-Fifth, 172 E. Chestnut, in the John
Hancock Center, 787-9596
Pizzeria Uno, 29 E. Ohio St., 321-1000
The Pump Room, in the Omni
Ambassador East Hotel, 266-0360
Su Casa, 49 E. Ontario St., 943-4041

SELECTED ATTRACTIONS:
Adler Planetarium, 1300 S. Lake Shore
Dr., 322-0300
Art Institute, Michigan Ave. at Adams St.,
443-3600
Brookfield Zoo, 31st St. & 1st Ave.,
Brookfield, (708) 485-0263
Field Museum of Natural History,
Roosevelt Rd. at S. Lake Shore Dr.,
922-9410
John G. Shedd Aquarium, 1200 S. Lake
Shore Dr., 939-2426
Lincoln Park Zoo, Fullerton Ave. & N.
Lake Shore Dr., 294-4660
The "Magnificent Mile" (shopping), N.
Michigan Ave.
Museum of Science & Industry, 57th St.
& S. Lake Shore Dr., 684-1414
Sears Tower, 233 S. Wacker Dr., 875-9696
Spertus Museum of Judaica, 618 S.
Michigan Ave., 922-9012

INFORMATION SOURCES:
Chicago Convention & Tourism Bureau,
Inc.
McCormick Place-on-the-Lake
2301 S. Lake Shore Dr.
Chicago, Illinois 60616
Tourism Info. (312) 280-5740
Chicagoland Chamber of Commerce
200 N. LaSalle St., 6th Floor
Chicago, Illinois 60601
(312) 580-6900

Cincinnati, Ohio

Population: (*1,452,645)
364,040 (1990C)
Altitude: 683 feet
Average Temp.: Jan., 35°F.; July, 78°F.
Telephone Area Code: 513
Time: 721-1700 **Weather:** 241-1010
Time Zone: Eastern

AIRPORT TRANSPORTATION:
Thirteen miles to downtown Cincinnati.

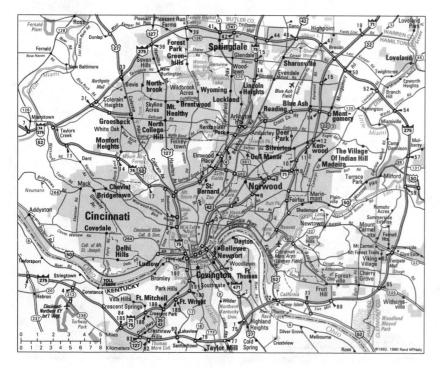

©1992, 1980 Rand McNally

SELECTED HOTELS:

Carrousel Inn, 8001 Reading Rd.,
821-5110

Cincinnati Marriott, 11320 Chester Rd.,
772-1720

The Clarion Hotel of Cincinnati, 141 W.
6th St., 352-2100

Harley Hotel of Cincinnati, 8020
Montgomery Rd., 793-4300

Holiday Inn, 3855 Hauk Rd., 563-8330

Howard Johnson Plaza, 11440 Chester
Rd., 771-3400

Hyatt Regency, 151 W. 5th St., 579-1234

Imperial House—West, 5510 Rybolt Rd.,
574-6000

Omni Netherland Plaza, 35 W. 5th St.,
421-9100

Terrace Hilton, 15 W. 6th St., 381-4000

The Vernon Manor Hotel, 400 Oak St.,
281-3300

The Westin Hotel, 5th & Vine, 621-7700

SELECTED RESTAURANTS:

Celestial, 1071 Celestial St., 241-4455

Gourmet Room, in the Terrace Hilton,
381-4000

La Normandie Grill, 118 E. 6th St.,
721-2761

Maisonette, 114 E. 6th St., 721-2260

The Orchids at the Palm Court, in the
Omni Netherland Plaza, 421-9100

The Palace Restaurant, in the
Cincinnatian Hotel, 601 Vine St.,
381-3000

Windjammer, 11330 Chester Rd.,
Sharonville, 771-3777

SELECTED ATTRACTIONS:

College Football Hall of Fame, 5440 Kings
Island Dr., 398-5410

Cincinnati Fire Museum, 315 W. Court St., 621-5553
Cincinnati Zoo, 3400 Vine St., 281-4700
Kings Island (theme park), Kings Island, 241-5600
Museum of Natural History, 1301 Western Ave., 287-7020
Showboat Majestic (live stage and musical shows), foot of Broadway, 241-6550 (Jan.—Oct.)
Tower Place at the Carew Tower (shopping), 4th & Race sts., 241-7700
William Howard Taft National Historic Site, 2038 Auburn Ave., 684-3262

INFORMATION SOURCES:
Greater Cincinnati Convention and Visitors Bureau
300 W. 6th St.
Cincinnati, Ohio 45202
(513) 621-2142
Greater Cincinnati Chamber of Commerce
300 Carew Tower
441 Vine St.
Cincinnati, Ohio 45202
(513) 579-3100

Cleveland, Ohio

Population: (*1,831,122) 505,616 (1990C)
Altitude: 570 to 1,050 feet
Average Temp.: Jan., 29°F.; July, 74°F.
Telephone Area Code: 216
Time: 931-1212 **Weather:** 267-3900
Time Zone: Eastern

AIRPORT TRANSPORTATION:
Twelve miles from Hopkins to downtown Cleveland.
Taxicab, train, and limousine bus service.

SELECTED HOTELS:
Cleveland Airport Marriott, 4277 W. 150th St., 252-5333
Cleveland Hilton South, 6200 Quarry Ln. at I-77 & Rockside Rd., 447-1300
Cleveland Marriott East, Park East Dr., I-271 & Chagrin Blvd., Beachwood, 464-5950
Harley Hotel East, 6051 SOM Center Rd., Willoughby, 944-4300
Harley Hotel West, 17000 Bagley Rd., 243-5200
Holiday Inn Lakeside City Center, 1111 Lakeside Ave., 241-5100
Ritz Carlton, 1515 W. 3rd St., 623-1300
Sheraton Cleveland City Center, 777 St. Clair Ave., 771-7600
Sheraton Hopkins Airport Hotel, 5300 Riverside Dr. at the airport, 267-1500
Stouffer's Tower City Plaza Hotel, 24 Public Sq., 696-5600

SELECTED RESTAURANTS:
Brasserie, in Stouffer's Tower City Plaza Hotel, 696-5600
Cafe Sausilito, 1301 E. 9th, 696-2233
Getty's at the Hanna, 1422 Euclid Ave., 771-1818
Sammy's, 1400 W. 10th St., 523-5560
Samurai Japanese Steak House, 23611 Chagrin Blvd., Beachwood, 464-7575
That Place on Bellflower, 11401 Bellflower Rd., 231-4469

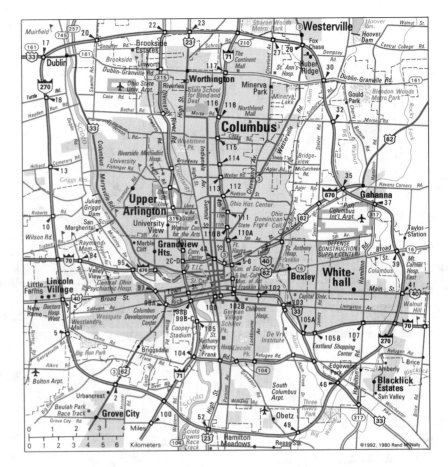

SELECTED ATTRACTIONS:
Cleveland Children's Museum, 10730 Euclid Ave., 791-7114
Cleveland Health Education Museum, 8911 Euclid Ave., 231-5010
Cleveland Museum of Art, 11150 E. Boulevard, on University Circle, 421-7340
Cleveland Museum of Natural History, University Circle, 231-4600
The Galleria (shopping), East 9th at St. Clair, 621-9999
Geauga Lake Amusement Park, 1060 Aurora Rd., Aurora, 562-7131
Metroparks Zoo, 3900 Brookside Park, 661-6500
NASA Visitor Center, 21000 Brookpark Rd., 433-4000
Sea World of Ohio, 1100 Sea World Dr., Aurora, 562-8101
Trolley Tours of Cleveland, 1831 Columbus Rd., 771-4484

INFORMATION SOURCES:
Convention and Visitors Bureau of Greater Cleveland
3100 Tower City Center
Cleveland, Ohio 44113
(216) 621-4110
Greater Cleveland Growth Association
200 Tower City Center
50 Public Square
Cleveland, Ohio 44113-2291
(216) 621-3300

Columbus, Ohio

Population: (*1,377,419) 632,910 (1990C)
Altitude: 685 to 893 feet
Average Temp.: Jan., 31°F.; July, 76°F.
Telephone Area Code: 614
Time: 281-8211 **Weather:** 231-5212
Time Zone: Eastern

AIRPORT TRANSPORTATION:
Eight miles to downtown Columbus.
Taxicab and limousine bus service.

SELECTED HOTELS:
Best Western North, 888 E. Dublin-Granville Rd., 888-8230
Columbus Marriott North, 6500 Doubletree Ave., 885-1885
Columbus Sheraton, 2124 S. Hamilton Rd., 861-7220
Concourse Hotel & Conference Center, 4300 International Gateway, 237-2515
Harley Hotel of Columbus, 1000 E. Dublin-Granville Rd., 888-4300
Hilton Inn North, 7007 N. High St., 436-0700
Holiday Inn Columbus Airport, 750 Stelzer/James Rd., 237-6360
Holiday Inn East, 4560 Hilton Corporate Dr., 868-1380
Holiday Inn on the Lane, 328 W. Lane Ave., 294-4848
Holiday Inn West, 2350 Westbelt Dr., 771-8999

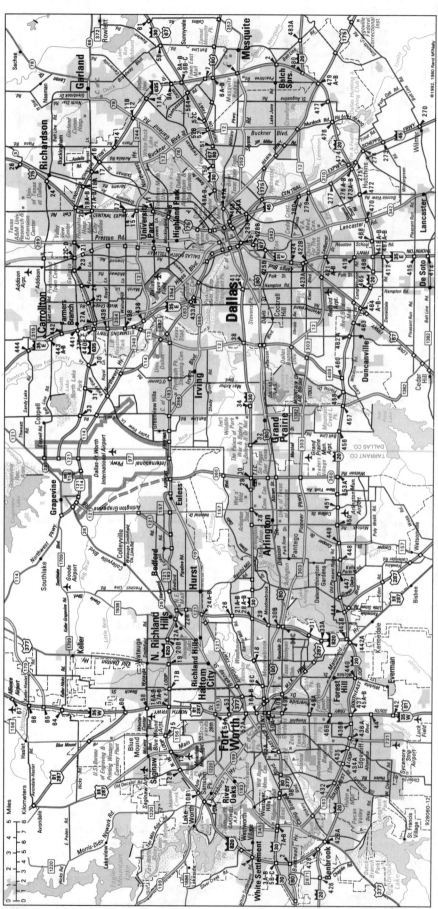

Hyatt Regency Columbus, 350 N. High
St., 463-1234
Radisson Airport Hotel, 1375 N. Cassady
Ave., 475-7551
Ramada University Hotel & Conference
Center, 3110 Olentangy River Rd.,
267-7461

SELECTED RESTAURANTS:
Allen O'Meara's, 89 Nationwide Blvd.,
621-2747
The Clarmont, 684 S. High St., 443-1125
Confluence Park, 679 W. Spring St.,
224-0000
Fifty-five on the Boulevard, 55
Nationwide Blvd., 228-5555
Jai Lai, 1421 Olentangy River Rd.,
421-7337
Kahiki, 3583 E. Broad St., 237-5425
One Nation, One Nationwide Plaza, 38th
Floor, 221-0151

SELECTED ATTRACTIONS:
Columbus Museum of Art, 480 E. Broad
St., 221-6801
Columbus Zoo, 9990 Riverside Dr.,
Powell, 645-3400
COSI (science & industry museum), 280
E. Broad St., 228-COSI
Fabulous Palace Theatre, 34 W. Broad St.,
469-1331
Franklin Park Conservatory, 1777 E.
Broad St., 645-7447
German Village Society, 588 S. 3rd St.,
221-8888
Ohio Historical Center/Ohio Village, 1982
Velma Ave., 297-2300
Ohio State University (tours), 1739 N.
High St., Room 249, 292-0428
Wexner Center for the Arts, N. High St. at
15th Ave., 292-0330
Wyandot Lake (water & amusement
park), 10101 Riverside Dr., 889-9283

INFORMATION SOURCES:
Greater Columbus Convention & Visitors
Bureau
10 W. Broad St., Suite 1300
Columbus, Ohio 43215
(614) 221-6623; (800) 234-2657
Columbus Area Chamber of Commerce
37 N. High St.
Columbus, Ohio 43215
(614) 221-1321

Dallas-Fort Worth, Texas

Population: (*3,885,415)
(Dallas) 1,006,877;
(Fort Worth) 447,619 (1990C)
Altitude: 463 to 750 feet
Average Temp.: Jan., 44°F.; July, 86°F.
Telephone Area Code: (Dallas) 214
(Fort Worth) 817
Time: 844-6611 **Weather:** (Dallas)
787-1111; (Fort Worth) 429-2631
Time Zone: Central

AIRPORT TRANSPORTATION:
About 17 miles to downtown Dallas or
Fort Worth.
Taxicab and limousine bus service.

SELECTED HOTELS: DALLAS
Adolphus Hotel, 1321 Commerce St.,
742-8200

Dallas Grand Hotel, 1914 Commerce St., 747-7000

Delux Inn, 3111 Stemmons Frwy., 637-0060

Fairmont Hotel, 1717 N. Akard St., 720-2020

Hiltop Hotel, 5600 N. Central Expwy., 827-4100

Hyatt Regency Dallas, 300 Reunion Blvd., 651-1234

Hyatt Regency—Dallas-Fort Worth Airport, International Pkwy., 453-8400

Le Baron Hotel, 1055 Regal Row, 634-8550

Loews Anatole Hotel, 2201 Stemmons Frwy., 748-1200

Mansion on Turtle Creek, 2821 Turtle Creek Blvd., 559-2100

The Marriott Mandalay, 221 E. Las Colinas Blvd., Irving, 556-0800

Plaza of the Americas Hotel, 650 N. Pearl St., 979-9000

Sheraton Park Central Hotel & Towers, 12720 Merit Dr., 385-3000

Stouffer Dallas Hotel, 2222 N. Stemmons Frwy., 631-2222

The Westin Hotel, 13340 Dallas Pkwy., 934-9494

SELECTED RESTAURANTS: DALLAS

Beau Nash, Hotel Crescent Court, 400 Crescent Ct., 871-3200

Il Sorrento, 8616 Turtle Creek Blvd., 352-8759

L'Entrecote, in the Loew's Anatole Hotel, 748-1200

Old Warsaw, 2610 Maple Ave., 528-0032

Plum Blossom, in the Loew's Anatole Hotel, 748-1200

The Pyramid Room, Fairmont Hotel, 720-2020

Routh Street Cafe, 3005 Routh St., 871-7161

650 North, Plaza of the Americas Hotel, 979-9000

SELECTED ATTRACTIONS: DALLAS

Biblical Arts Center, 7500 Park Lane, 691-4661

Boardwalk (family fun park), I-30 & Beltline, Grand Prairie, 642-7900

Dallas Arboretum & Botanical Gardens, 8617 Garland Rd., 327-8263

Dallas Museum of Art, 1717 N. Harwood St., 922-1200

Dallas Zoo, in Marsalis Park, 621 E. Clarendon, 946-5154

Fair Park (museums), 565-9931

The Meadows Museum (15th-20th century Spanish art), Meadows School of the Arts, Southern Methodist University campus, 692-2516

Old City Park (museum village), 1717 Gano, 421-5141

The Sixth Floor (President John F. Kennedy educational exhibit), Houston & Elm St., 653-6666

West End Historical District, on the west end of Dallas' central business district

INFORMATION SOURCES: DALLAS

Dallas Convention & Visitors Bureau
1201 Elm St., Suite 2000
Dallas, Texas 75270
(214) 746-6677

Dallas Chamber of Commerce

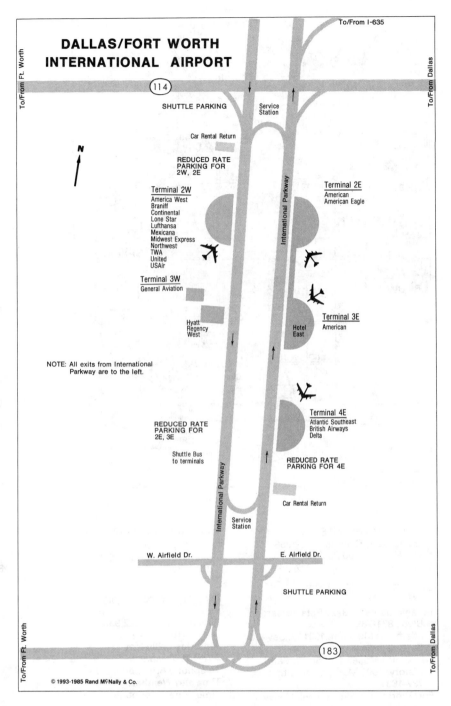

DALLAS/FORT WORTH INTERNATIONAL AIRPORT

To/From Ft. Worth

To/From Dallas

114

To/From I-635

SHUTTLE PARKING

Service Station

Car Rental Return

REDUCED RATE PARKING FOR 2W, 2E

N

Terminal 2W
America West
Braniff
Continental
Lone Star
Lufthansa
Mexicana
Midwest Express
Northwest
TWA
United
USAir

Terminal 2E
American
American Eagle

International Parkway

Terminal 3W
General Aviation

Hyatt Regency West

Hotel East

Terminal 3E
American

NOTE: All exits from International Parkway are to the left.

REDUCED RATE PARKING FOR 2E, 3E

Terminal 4E
Atlantic Southeast
British Airways
Delta

Shuttle Bus to terminals

REDUCED RATE PARKING FOR 4E

Car Rental Return

International Parkway

Service Station

W. Airfield Dr.

E. Airfield Dr.

SHUTTLE PARKING

To/From Ft. Worth

To/From Dallas

183

© 1993-1985 Rand McNally & Co.

Information Department
1201 Elm St., Suite 2000
Dallas, Texas 75270
(214) 746-6700

SELECTED HOTELS: FORT WORTH

Chisholm Hotel, 600 Commerce St., 332-6900

Clarion Hotel Fort Worth, 2000 Beach St., 534-4801

Green Oaks Inn, 6901 W. Frwy., 738-7311

HoJo Inn, I-35 W. South at Seminary Exit, 923-8281

La Quinta—Fort Worth West, 7888 I-30 W. at Cherry Ln., 246-5511

Park Central Hotel, 1010 Houston St., 336-2011

Radisson Plaza, 815 Main St., 870-2100

Ramada Hotel, 1701 Commerce St., 335-7000

Ramada Inn, Jct. TX 183 & I-820, 284-9461

The Worthington, 200 Main St., 870-1000

SELECTED RESTAURANTS: FORT WORTH

The Balcony, 6100 Camp Bowie Blvd., 731-3719

The Cattle Drive, 1900 Ben St., 534-4908

Crystal Cactus, in the Radisson Plaza, 870-2100

Juanita's, 115 W. 2nd St., 335-1777
The Keg, 1309 Calhoun, 332-1288
Mac's House, 2400 Park Hill Dr., 921-4682
Tours, 3500 W. 7th, 870-1672

SELECTED ATTRACTIONS:
FORT WORTH
Botanic Gardens, 3220 Botanic Garden
  Blvd., 871-7686
Cattleman's Museum, 1301 W. Seventh
  St., 332-7064
Fort Worth Museum of Science and
  History, 1501 Montgomery St.,
  732-1631
Fort Worth Zoo, 1989 Colonial Pkwy.,
  871-7050 or -7051
Log Cabin Village, on University across
  from the zoo, 926-5881
Modern Art Museum of Fort Worth, 1309
  Montgomery St., 738-9215
Sid Richardson Collection of Western Art,
  309 Main St., 332-6554
Tarantula Railroad, 2318 8th Ave.,
  763-8297
Thistle Hill (cattle baron mansion), 1509
  Pennsylvania Ave., 336-1212

INFORMATION SOURCES:
FORT WORTH
Fort Worth Convention & Visitors Bureau
  415 Throckmorton St.
  Fort Worth, Texas 76102
  (817) 336-8791; (800) 433-5747

Fort Worth Chamber of Commerce
  777 Taylor, Suite 900
  Fort Worth, Texas 76102
  (817) 336-2491

Denver, Colorado

Population: (*1,622,980)
  467,610 (1990C)
Altitude: 5,280 feet
Average Temp.: Jan., 31°F.; July, 74°F.
Telephone Area Code: 303
Time and Weather: 1-976-1311
Time Zone: Mountain

AIRPORT TRANSPORTATION:
Seven miles to downtown Denver.
Bus, taxicab, and limousine bus service.

SELECTED HOTELS:
Brown Palace Hotel, 321 17th St.,
  297-3111
Burnsley Hotel, 1000 Grant St., 830-1000
Denver Marriott City Center, 1701
  California, 297-1300
Denver Marriott Southeast, 6363 E.
  Hampden Ave., 758-7000
Holiday Inn Denver Southeast, I-225 &
  South Parker Rd., 695-1700
Hotel Denver Downtown, 1450 Glenarm
  Pl., 573-1450
Hyatt Regency Denver, 1750 Welton St.,
  295-1200

Oxford Hotel, 1600 17th St., 628-5400
Radisson Hotel Denver, 1550 Court Pl.,
  893-3333
Red Lion Hotel, 3203 Quebec St.,
  321-3333
Stapleton Plaza Hotel and Fitness Center,
  3333 Quebec St., 321-3500

SELECTED RESTAURANTS:
Ellyngton's, at the Brown Palace Hotel,
  297-3111
Marlowe's, 511 16th St., 595-3700
Normandy French Restaurant, 1515
  Madison St., 321-3311
Palace Arms, at the Brown Palace Hotel,
  297-3111
Racines, 850 Bannock, 595-0418
Strings, 1700 Humboldt, 831-7310
Tante Louise, 4900 E. Colfax Ave.,
  355-4488

SELECTED ATTRACTIONS:
Buffalo Bill's Grave and Museum, top of
  Lookout Mountain, 526-0747
Children's Museum of Denver, 2121
  Children's Museum Dr., 433-7444
Colorado Railroad Museum, 17155 W.
  44th Ave., Golden, 279-4591
Coors Brewing Company (tours), 13th &
  Ford sts., Golden, 277-BEER
Denver Center for the Performing Arts,
  14th & Curtis sts., 893-4100
Denver Museum of Natural History, 22nd
  & Colorado Blvd. in City Park, 322-7009
Denver's Zoo, 23rd and Steele St., in City
  Park, 331-4110
Larimer Square (restored Victorian-era
  section of Denver), 1400 block of
  Larimer
Molly Brown House Museum, 1340
  Pennsylvania, 832-4092
United States Mint, 320 W. Colfax Ave.,
  844-3582

INFORMATION SOURCES:
Denver Metro Convention and Visitors
Bureau
  225 W. Colfax Ave.
  Denver, Colorado 80202
  (303) 892-1112
Greater Denver Chamber of Commerce
  1445 Market St.
  Denver, Colorado 80202
  (303) 534-8500

Detroit, Michigan

Population: (*4,382,299)
  1,027,974 (1990C)
Altitude: 573 to 672 feet
Average Temp.: Jan., 26°F.; July, 73°F.
Telephone Area Code: 313
Time: 472-1212 **Weather:** 1-976-1000
Time Zone: Eastern

AIRPORT TRANSPORTATION:
Nineteen miles from Metropolitan Airport
  to downtown Detroit.
Taxicab and limousine bus service.

SELECTED HOTELS:
Holiday Inn Detroit Metro Airport, 31200
  Industrial Expwy., 728-2800
Hotel St. Regis, 3071 W. Grand Blvd.,
  873-3000
Hyatt Regency Dearborn, Fairlane Town
  Center Dr., Dearborn, 593-1234

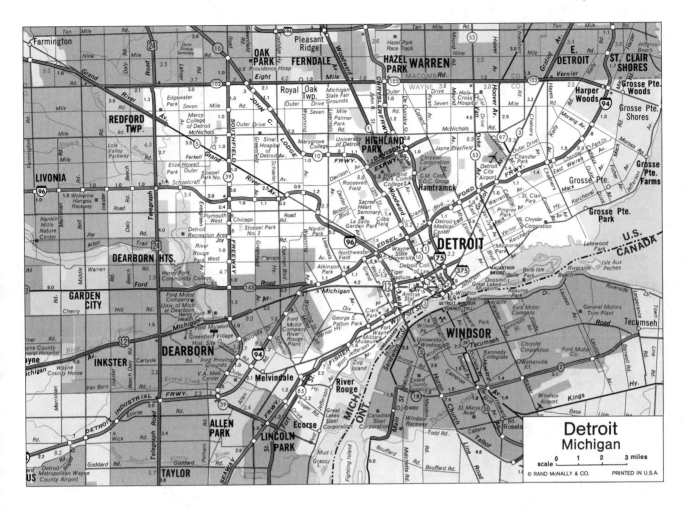

Detroit
Michigan

scale 0   1   2   3 miles

© RAND McNALLY & CO.   PRINTED IN U.S.A.

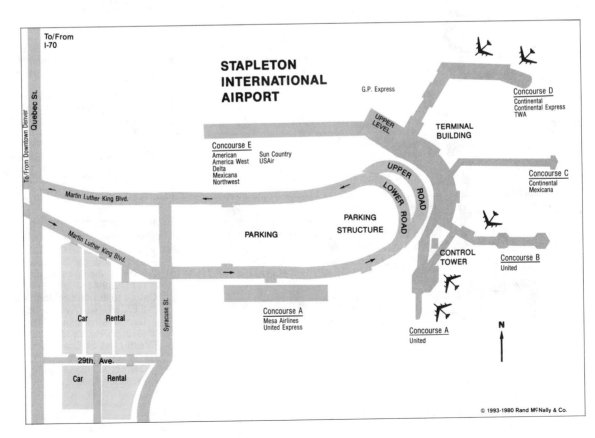

To/From
I-70

STAPLETON
INTERNATIONAL
AIRPORT

G.P. Express

Concourse D
Continental
Continental Express
TWA

TERMINAL
BUILDING

UPPER
LEVEL

Concourse E
American   Sun Country
America West   USAir
Delta
Mexicana
Northwest

Concourse C
Continental
Mexicana

UPPER
ROAD

LOWER
ROAD

Martin Luther King Blvd.

PARKING
STRUCTURE

PARKING

CONTROL
TOWER

Concourse B
United

Car   Rental

Concourse A
Mesa Airlines
United Express

Concourse A
United

Martin Luther King Blvd.

29th. Ave.

Car   Rental

N

© 1993-1980 Rand McNally & Co.

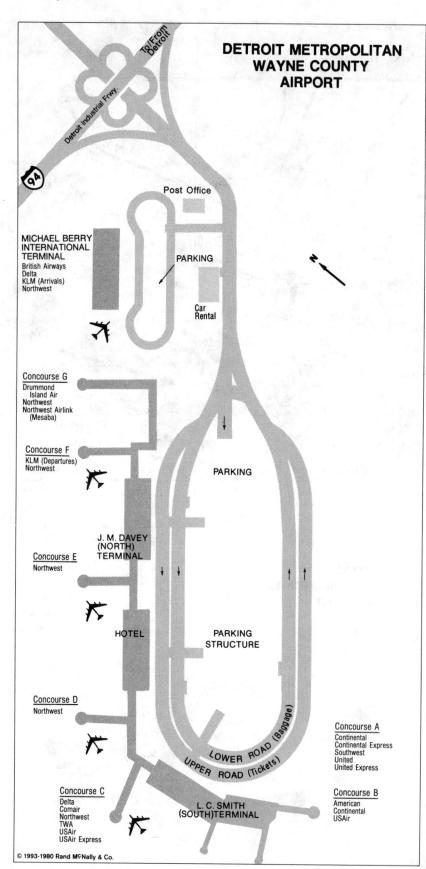

DETROIT METROPOLITAN WAYNE COUNTY AIRPORT

Northfield Hilton, 5500 Crooks Rd., Troy, 879-2100
Omni International, 333 E. Jefferson, 222-7700
Radisson Hotel Pontchartrain, 2 Washington Blvd., 965-0200
Royce Hotel, 31500 Wick Rd., Romulus, 292-3400
Sheraton Southfield, 16400 J.L. Hudson Dr., Southfield, 559-6500
The Westin Hotel, Renaissance Center, 568-8000

SELECTED RESTAURANTS:
Captains, 260 Schweizer Place, 568-1862
Carl's Chop House, 3026 Grand River Ave., 833-0700
Caucus Club, 150 W. Congress St., 965-4970
Charley's Crab, 5498 Crooks Rd., Troy, 879-2060
The Golden Mushroom, 18100 W. 10 Mile Rd., Southfield, 559-4230
Joe Muer's Sea Food, 2000 Gratiot Ave., 567-1088
London Chop House, 155 W. Congress St., 962-6735
Mario's Restaurant, 4222 2nd Ave., 832-1616
Opus One, 565 E. Larned, 961-7766
St. Regis Restaurant, in the St. Regis Hotel, 3071 W. Grand Blvd., 873-3000
The Summit, in the Westin Hotel, 568-8000
Van Dyke Place, 649 Van Dyke Ave., 821-2620

SELECTED ATTRACTIONS:
Cranbrook Academy of Art Museum, 500 Lone Pine Rd., Bloomfield Hills, 645-3323
Detroit Institute of Arts, 5200 Woodward Ave., 833-7900
Detroit Zoological Park, 8450 W. Ten Mile Rd., Royal Oak, 398-0903
Edsel & Eleanor Ford House, 1100 Lakeshore Rd., Grosse Pointe Shores, 884-4222 or 884-3400
Graystone International Jazz Museum, 1521 Broadway, 963-3813
Henry Ford Estate-Fair Lane, University of Michigan—Dearborn campus, 4901 Evergreen, Dearborn, 593-5590
Henry Ford Museum & Greenfield Village, 20900 Oakwood Blvd., Dearborn, 271-1620
Motown Museum, 2648 W. Grand Blvd., 867-0991

INFORMATION SOURCES:
Metropolitan Detroit Convention & Visitors Bureau
  Suite 1950, 100 Renaissance Center
  Detroit, Michigan 48243
  (313) 259-4333
Detroit Visitor Information Center
  2 E. Jefferson Ave., Civic Center Area
  Detroit, Michigan 48226
  (313) 567-1170; (800) 338-7648
Greater Detroit Chamber of Commerce
  600 W. Lafayette
  Detroit, Michigan 48226
  (313) 964-4000

Edmonton, Alberta, Canada

Population: (*785,465)
   573,982 (1986C)
Altitude: 2,192 feet
Average Temp.: Jan., 6°F.; July, 62°F.
Telephone Area Code: 403
Time: 421-1111   **Weather:** 468-4940
Time Zone: Mountain

AIRPORT TRANSPORTATION:

Eighteen miles from Edmonton
  International to downtown
  Edmonton; three miles from Edmonton
  Municipal to downtown Edmonton.
Taxicab; Airporter bus to and from
  Edmonton International; shuttle
  service between Edmonton
  International and Edmonton Municipal.

SELECTED HOTELS:

Centre Suite Hotel, 10222 102nd St.,
  429-3900
Convention Inn, 4404 Calgary Trail
  Northbound, 434-6415
Edmonton Hilton, 10235 101st St.,
  428-7111
Edmonton Inn, 11830 Kingsway Ave.,
  454-9521
Holiday Inn Crowne Plaza, 1011 Bellamy
  Hill, 428-6611
Nisku Inn, 4th St. & 11th Ave., across
  from International Airport, 955-7744
Radisson, 10010 104th St., 423-2450
Ramada Renaissance, 10155 105th St.,
  423-4811
The Westin Hotel—Edmonton, 10135
  100th St., 426-3636
Westwood Inn Best Western, 18035
  Stony Plain Rd., 483-7770

SELECTED RESTAURANTS:

Boccalino Swiss Italian Pasta Bistro,
  10525 Jasper Ave., 426-7313
Boulevard Café, in the Ramada
  Renaissance Hotel, 423-4811
The Carvery, in the Westin Hotel,
  426-3636
Claude's, 10112 107th St., 429-2900
Japanese Village, 10126 100th St.,
  422-6083
La Ronde, in the Holiday Inn Crowne
  Plaza, 428-6611
The Mill Gasthaus, 8109 101st St.,
  432-1838
Pacific Fish Company, 1638 Bourbon St.,
  in West Edmonton Mall, 444-1905
The Tin Palace, 11830 Jasper Ave.,
  488-6582
Top of the Inn, in the Convention Inn
  Hotel, 434-6415

SELECTED ATTRACTIONS:

Alberta Railroad Museum, 24215 34th St.
  NE, 472-6229
Edmonton Space & Science Centre,
  11211 142nd St., 451-7722
Elk Island National Park, 35 mi. east on
  Hwy. 16, 992-6380
Fort Edmonton Park, Whitemud & Fox
  drs., 428-2992
Muttart Conservatory, 98th Ave. & 96A
  St., 428-5226
Provincial Museum of Alberta, 128th St.
  & 102nd Ave., 427-1786
Ukrainian Cultural Heritage Museum, 30
  mi. east on Hwy. 16, 662-3640

West Edmonton Mall/Canada
  Fantasyland, 8770 170th St., 444-5200
World Wildlife Learning Center, 8¼ mi.
  north of Bon Accord on Lily Lake Rd.,
  921-2294

INFORMATION SOURCES:

Edmonton Convention & Tourism
Authority
  9797 Jasper Ave., No. 104
  Edmonton, Alberta, Canada T5J 1N9
  (403) 426-4715
Edmonton Chamber of Commerce
  10123 99th St., Suite 600
  Edmonton, Alberta, Canada T5J 3G9
  (403) 426-4620

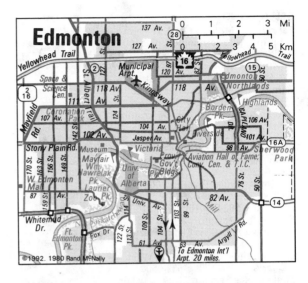

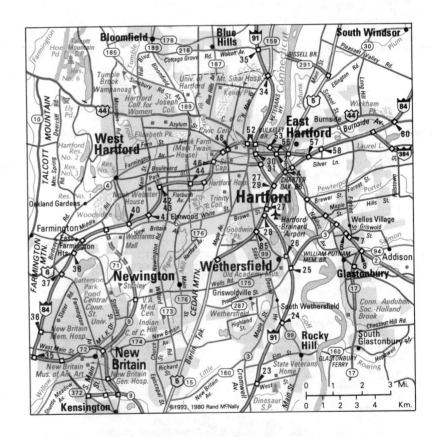

Hartford, Connecticut

Population: (*767,841)
   139,739 (1990C)
Altitude: 10 to 290 feet
Average Temp.: Jan., 28°F.; July, 74°F.
Telephone Area Code: 203
Time: 524-8123   **Weather:** 1-936-1212
Time Zone: Eastern

AIRPORT TRANSPORTATION:

Thirteen miles to downtown Hartford.
Taxicab and limousine bus service.

SELECTED HOTELS:
Farmington Marriott Hotel, 15 Farm
    Springs Rd., Farmington, 678-1000
Holiday Inn—Downtown, 50 Morgan St.,
    549-2400
J.P. Morgan Hotel, 1 Haynes St., 246-7500
Ramada Hotel, 100 E. River Dr., E.
    Hartford, 528-9703
Sheraton Hartford Hotel, 315 Trumbull
    St., 728-5151

SELECTED RESTAURANTS:
Avon Old Farms Inn, U.S. 44 & CT 10,
    677-2818
The Blacksmith's Tavern, 2300 Main St.,
    Glastonbury, 659-0366
Carbone's, 588 Franklin Ave., 296-9646
Frank's, 185 Asylum, 527-9291
Gaetano's, One Civic Center Plaza,
    249-1629
L'Americain, Hartford Square West,
    Charter Oak & Willis, 522-6500
Market Place Restaurant, 39 New London
    Tpke., Glastonbury, 633-3832
Parson's Daughter, 2 Hopewell Rd., S.
    Glastonbury, 633-8698

SELECTED ATTRACTIONS:
Bushnell Park Carousel Society
    (hand-carved carousel), 246-7739
Butler-McCook Homestead (1782), 396
    Main St., 522-1806
Center Church & The Ancient Burying
    Ground, Main & Gold sts., 249-5631
Elizabeth Park (rose garden), Prospect
    Ave. & Asylum St.
Historical Museum of Medicine &
    Dentistry, 230 Scarborough St.,
    236-5613
Mark Twain & Harriet Beecher Stowe
    houses, Visitor Center, 77 Forest St.,
    525-9317
Old State House, 800 Main St., 522-6766
Raymond E. Baldwin Museum of
    Connecticut History, 231 Capitol Ave.,
    566-3056
Wadsworth Atheneum (first American
    public art museum), 600 Main St.,
    278-2670

INFORMATION SOURCES:
Greater Hartford Convention & Visitors
Bureau, Inc.
    One Civic Center Plaza
    Hartford, Connecticut 06103
    (203) 728-6789
    Visitor Information: (203) 541-6440
Old Statehouse Information Center
    800 Main St.
    (brochures only; self-service)
Greater Hartford Chamber of Commerce
    250 Constitution Plaza
    Hartford, Connecticut 06103
    (203) 525-4451

Honolulu, Hawaii

Population: (*836,231)
    365,272 (1990C)
Altitude: 18 feet
Average Temp.: Jan., 72°F.; July, 80°F.
Telephone Area Code: 808
Time: 983-3211   **Weather:** 836-0121
Time Zone: Hawaiian (Two hours earlier
    than Pacific standard time)

AIRPORT TRANSPORTATION:
Nine miles to Waikiki.
Taxicab and limousine bus service.

SELECTED HOTELS:
Halekulani, 2199 Kalia Rd., 923-2311
Hawaiian Regent, 2552 Kalakaua Ave.,
    922-6611
Hilton Hawaiian Village, 2005 Kalia Rd.,
    949-4321
Hyatt Regency Waikiki, 2424 Kalakaua
    Ave., 923-1234
The Ilikai, 1777 Ala Moana Blvd.,
    949-3811
Kahala Hilton, 5000 Kahala Ave.,
    734-2211
Moana Surfrider, 2365 Kalakaua Ave.,
    922-3111
Outrigger Waikiki Hotel, 2335 Kalakaua
    Ave., 923-0711
Queen Kapiolani, 150 Kapahulu Ave.,
    922-1941
Royal Hawaiian, 2259 Kalakaua Ave.,
    923-7311
Sheraton Waikiki, 2255 Kalakaua Ave.,
    922-4422

SELECTED RESTAURANTS:
Furusato, 2500 Kalakaua Ave., 922-5502
Golden Dragon Room (Chinese), in the
    Hilton Hawaiian Village, 949-4321
The Hanohano Room, in the Sheraton
    Waikiki, 922-4422
Maile Room, in the Kahala Hilton,
    734-2211
Michel's, in the Colony Surf Hotel, 2895
    Kalakaua Ave., 923-6552
The Plantation Cafe, in the Ala Moana
    Hotel, 410 Atkinson Dr., 955-4811

The Secret, in the Hawaiian Regent Hotel,
    922-6611
The Willows, 901 Hausten St., 946-4808

SELECTED ATTRACTIONS:
Bernice P. Bishop Museum &
    Planetarium, 1525 Bernice St., 847-3511
Honolulu Zoo, 151 Kapahulu Ave.,
    971-7171
National Memorial Cemetery of the
    Pacific, overlooking Honolulu, 541-1430
Pearl Harbor/The U.S.S. Arizona
    Memorial, 422-0561
Polynesian Cultural Center, 55-370 Kam
    Hwy., Laie, 293-3333
Sea Life Park, Makapuu Point, 259-7933
Waikiki Aquarium, 2777 Kalakaua Ave.,
    923-9741
Waimea Falls Park, 59-864 Kam Hwy. on
    the north shore of Oahu, 638-8511

INFORMATION SOURCES:
Hawaii Visitors Bureau
    Meetings and Conventions Office
    2270 Kalakaua Ave., 8th Floor
    Honolulu, Hawaii 96815
    (808) 923-1811
The Chamber of Commerce of Hawaii
    735 Bishop St., Suite 220
    Honolulu, Hawaii 96813
    (808) 522-8800

Houston, Texas

Population: (*3,301,937)
    1,630,553 (1990C)
Altitude: Sea level to 50 feet
Average Temp.: Jan., 59°F.; July, 82°F.
Telephone Area Code: 713

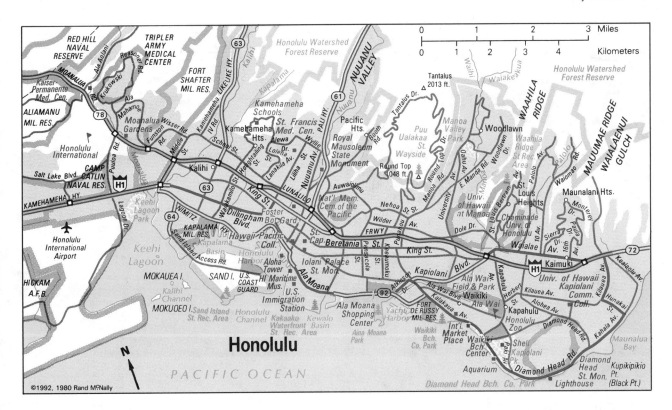

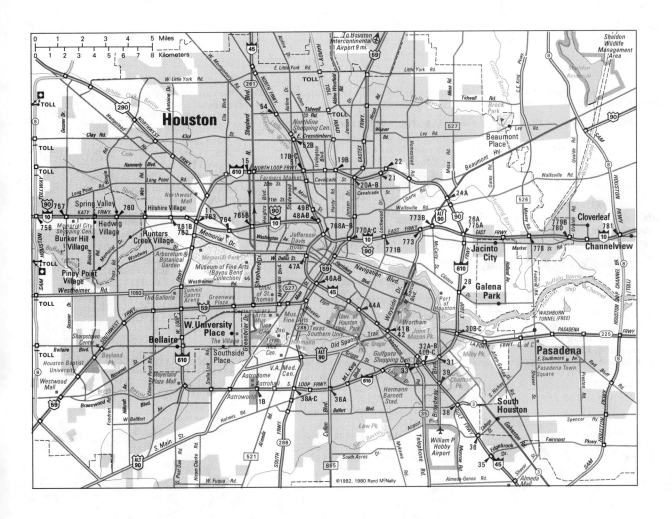

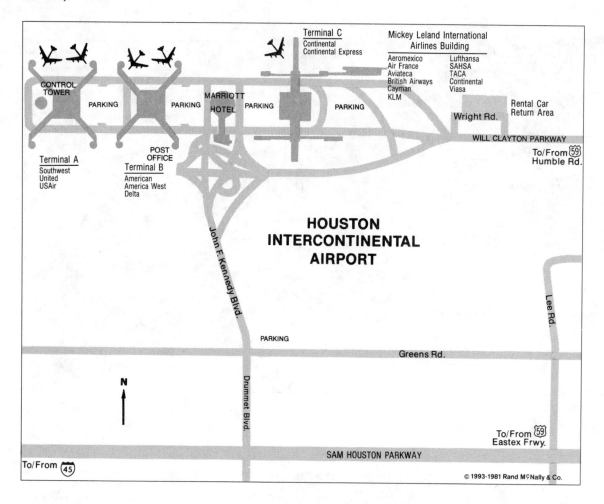

Time: 222-8463   **Weather:** 529-4444
Time Zone: Central

AIRPORT TRANSPORTATION:
Twenty miles from Intercontinental
Airport to downtown Houston; 10
miles from Hobby Airport to downtown.
Taxicab, limousine, coach bus, and
helicopter service.

SELECTED HOTELS:
Adam's Mark, 2900 Briarpark at
Westheimer, 978-7400
Days Inn Downtown, 801 Calhoun St.,
659-2222
Doubletree at Allen Center, 400 Dallas St.,
759-0202
Four Seasons Hotel, Houston Center,
1300 Lamar, 650-1300
Guest Quarters—Galleria West, 5353
Westheimer, 961-9000
Houston Marriott Astrodome, 2100 S.
Braeswood, 797-9000
Hyatt Regency Houston, 1200 Louisiana
St., 654-1234
Omni Houston, 4 Riverway Dr., 871-8181
The Sheraton Grand, 2525 W. Loop
South, 961-3000
Stouffer Presidente Hotel, Southwest
Frwy. at Edloe, 629-1200
The Westin Galleria, 5060 W. Alabama
St., 960-8100
The Westin Oaks Hotel, 5011 Westheimer
Rd., 623-4300

The Wyndham Warwick Hotel, 5701 Main
St., 526-1991

SELECTED RESTAURANTS:
Brennan's, 3300 Smith St., 522-9711
De Ville, in the Four Seasons Hotel,
Houston Center, 650-1300
The Great Caruso, 10001 Westheimer
Rd., 780-4900
Harry's Kenya, 1160 Smith St., 650-1980
La Tour d'Argent, 2011 Ella Blvd.,
864-9864
Maxim's, 3755 Richmond, 877-8899
The Rivoli, 5636 Richmond Ave.,
789-1900
Tony's, 1801 Post Oak Blvd., 622-6778
Vargo's, 2401 Fondren Rd., 782-3888

SELECTED ATTRACTIONS:
Battleship *Texas,* in San Jacinto
Battleground State Historic Park, 22
miles east of Houston off Texas
Highway 225 East, 479-2411
Bayou Bend (residence/museum of
American decorative arts—under
renovation, will reopen in fall, 1993;
gardens), 1 Westcott, (tours of
residence by reservation only)
529-8773
Houston Zoo, in Hermann Park, off the
6300 block of Fannin St., 525-3300
Museum of Fine Arts, 1001 Bissonnet,
639-7375

San Jacinto Monument & Museum of
History, in San Jacinto Battleground
State Historic Park, 22 miles east of
Houston off Texas Highway 225 East,
479-2421
Space Center Houston, 1601 NASA Rd. 1,
244-2100
The Galleria (shopping, dining,
entertainment), Loop 610 at
Westheimer, 621-1907

INFORMATION SOURCES:
Greater Houston Convention & Visitors
Bureau
3300 Main St.
Houston, Texas 77002-9396
(713) 523-5050; (800) 231-7799
Greater Houston Partnership
1100 Milam Building, 25th Floor
Houston, Texas 77002
(713) 651-1313

Indianapolis, Indiana

Population: (*1,249,822)
731,327 (1990C)
Altitude: 717 feet
Average Temp.: Jan., 29°F.; July, 76°F.
Telephone Area Code: 317
Time and Weather: 222-2222
Time Zone: Eastern standard all year

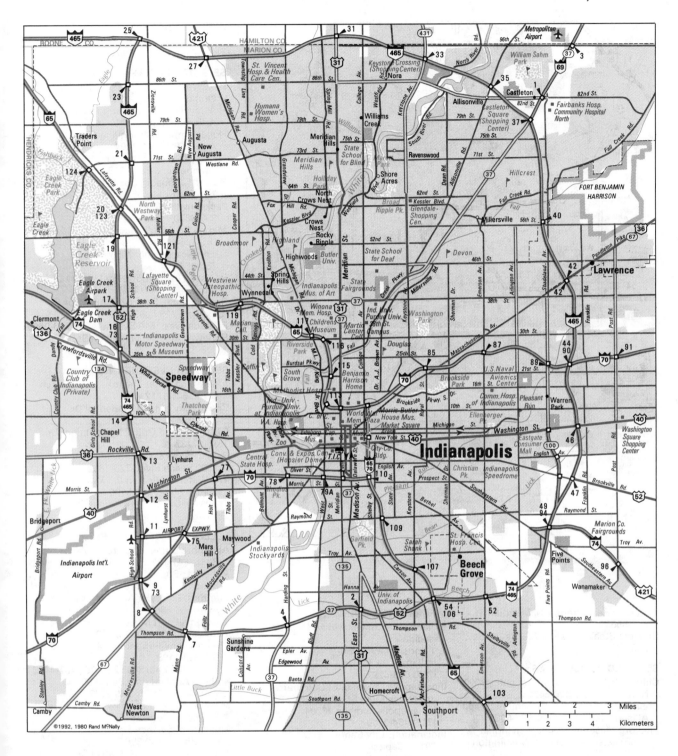

AIRPORT TRANSPORTATION:
Eight miles to downtown Indianapolis.
Taxicab and limousine bus service.

SELECTED HOTELS:
Adam's Mark Hotel, 2544 Executive
 Drive, 248-2481
Courtyard by Mariott, 501 W. Washington
 St., 635-4443
Holiday Inn Airport, 2501 S. High School
 Rd., 244-6861

Holiday Inn Union Station, 123 W.
 Louisiana, 631-2221
Hyatt Regency Indianapolis, 1 S. Capitol
 Ave., 632-1234
Indianapolis Airport Hilton, 2500 S. High
 School Rd., 244-3361
Indianapolis Hilton at the Circle, 31 W.
 Ohio St., 635-2000
Indianapolis Marriott Hotel, 7202 E. 21st
 St., 352-1231
Indianapolis Motor Speedway Motel,
 4400 W. 16th St., 241-2500

Omni Severin Hotel, 40 W. Jackson Pl.,
 634-6664
Westin Hotel, 50 S. Capitol Ave., 262-8100

SELECTED RESTAURANTS:
Chanteclair Sur Le Toit, Holiday Inn
 Airport, 244-6861
Chez Jean Restaurant Francais & Inn,
 9027 S. State Rd. 67, Camby, 831-0870
The City Taproom & Grille, 28 S.
 Pennsylvania St., 637-1334
The Eagle's Nest Restaurant, in the Hyatt
 Regency Indianapolis, 632-1234

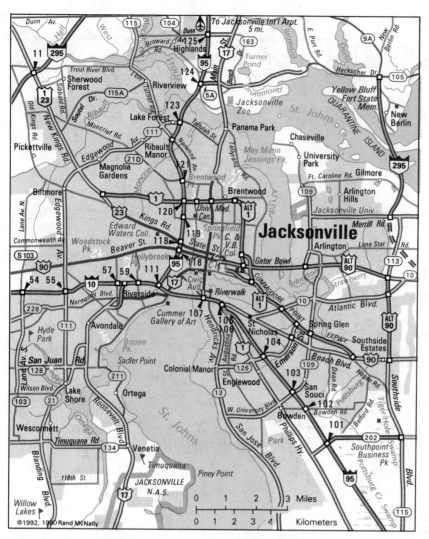

AIRPORT TRANSPORTATION:
Fifteen miles to downtown Jacksonville. Taxicab and limousine service.

SELECTED HOTELS:
Best Western Orange Park, 300 Park Ave., Orange Park, 264-1211
Holiday Inn and Conference Center, 5865 Arlington Expwy., 724-3410
Holiday Inn—Jacksonville Airport, I-95 at Airport Rd., 741-4404
Holiday Inn—Orange Park, 150 Park Ave., Orange Park, 264-9513
Howard Johnson's, 6545 Ramona Blvd., 781-1940
Jacksonville Hotel, 565 S. Main St., 398-8800
Marina St. John's Place, 1515 Prudential Dr., 396-5100
Omni Jacksonville, 245 Water St., 355-6664
Radisson Inn Airport Hotel, 14000 Dixie Clipper Dr., 741-4747
Ramada Inn—East, 6237 Arlington Expwy., 725-5093
Ramada West, 510 S. Lane Ave., 786-0500
Rodeway Inn, 3233 Emerson St., 398-3331

SELECTED RESTAURANTS:
Alhambra Dinner Theatre, 12000 Beach Blvd., 641-1212
Crawdaddy's, 1643 Prudential Dr., 396-3546
L & N Seafood Grill, 2 Independent Dr., 358-7737
Patti's Italian & American Restaurant, 7300 Beach Blvd., 725-1662
The Water Garden, in the Marina St. John's Place, 396-5100
The Wine Cellar, 1314 Prudential Dr., 398-8989

SELECTED ATTRACTIONS:
Alhambra Dinner Theatre, 12000 Beach Blvd., 641-1212
Anheuser-Busch Brewery (tours), 111 Busch Dr., 751-8116
Cummer Gallery of Art, 829 Riverside Ave., 356-6857
Jacksonville Art Museum, 4160 Boulevard Center Dr., 398-8336
The Jacksonville Landing, 2 Independent Dr., 353-1188
Jacksonville Zoo, 8605 Zoo Rd., 757-4463
Kathryn Abbel Hanna Park, 500 Wonderwood Dr., 249-2317
Mayport Naval Station (weekend tours), Mayport, 270-5226
Museum of Science & History, 1025 Museum Circle, 396-7062
The Riverwalk, 851 N. Market St., 396-4900

INFORMATION SOURCES:
Jacksonville & the Beaches Convention & Visitors Bureau
6 E. Bay St., Suite 200
Jacksonville, Florida 32202
(904) 798-9148
Jacksonville Chamber of Commerce
3 Independent Dr.
Jacksonville, Florida 32202
(904) 353-0300

King Cole Restaurant, 7 N. Meridian St., 638-5588
St. Elmo Steak House, 127 S. Illinois St., 635-0636

SELECTED ATTRACTIONS:
Children's Museum of Indianapolis, 3000 N. Meridian St., 924-5431
Conner Prairie Pioneer Settlement, 13400 Allisonville Rd., Fishers, 776-6000
Eiteljorg Museum of American Indian and Western Art, 500 W. Washington St., 636-9378
Indiana Dinner Train, 262-3333
Indiana State Museum & Historic Sites, 202 N. Alabama St., 232-1637
Indianapolis "500" Motor Speedway and Hall of Fame Museum, 4790 W. 16th St., 248-6746
Indianapolis Museum of Art, 1200 W. 38th St., 923-1331
Indianapolis Zoo, 1200 W. Washington St., 630-2030
Madame Walker Urban Life Center, 617 Indiana Ave., 236-2099
Union Station Festival Marketplace, 39 W. Jackson Place, 267-0701

INFORMATION SOURCES:
Indianapolis Convention and Visitors Association
1 Hoosier Dome, Suite 100
Indianapolis, Indiana 46225
(317) 639-4282
Indianapolis Chamber of Commerce
320 N. Meridian St., Suite 928
Indianapolis, Indiana 46204
(317) 464-2200
Indianapolis City Center
201 S. Capitol
Indianapolis, Indiana 46225
(317) 237-5200

Jacksonville, Florida

Population: (*906,727)
635,230 (1990C)
Altitude: 19 feet
Average Temp.: Jan., 55°F.; July, 80°F.
Telephone Area Code: 904
Time: (none) **Weather:** 741-4311
Time Zone: Eastern

Kansas City, Missouri

Population: (*1,566,280)
435,146 (1990C)
Altitude: 800 feet
Average Temp.: Jan., 30°F.; July, 81°F.
Telephone Area Code: 816
Time: 844-1212   **Weather:** 471-4840
Time Zone: Central

AIRPORT TRANSPORTATION:
Eighteen miles to downtown Kansas City.
Taxicab and limousine bus service.

SELECTED HOTELS:
Adam's Mark of Kansas City, 9103 E. 39th St., 737-0200
Airport Hilton Plaza Inn, I-29 & 112th St. NW., 891-8900
Allis Plaza Hotel, 200 W. 12th St., 421-6800
Americana Hotel, 1301 Wyandotte St., 221-8800
Hilton Plaza Inn, 45th & Main sts., 753-7400
Holiday Inn Crowne Plaza, 4445 Main St., 531-3000
Hyatt Regency Crown Center, 2345 McGee, 421-1234
Kansas City Airport Marriott, 775 Brasilia, 464-2200
Park Place Hotel, 1601 N. Universal Ave., 483-9900
Ritz Carlton Hotel, Kansas City, 401 Ward Pkwy., 756-1500
Westin Crown Center, 1 Pershing Rd., 474-4400

SELECTED RESTAURANTS:
American Restaurant, 200 E. 25th St., 426-1133
The Golden Ox, 1600 Genessee, 842-2866
The Hereford House, 20th & Main, 842-1080
Jasper's Restaurant, 405 W. 75th St., 363-3003
La Mediterranee, 4742 Pennsylvania St., 561-2916
Peppercorn Duck Club, in the Hyatt Regency Crown Center, 421-1234
Plaza III, 4749 Pennsylvania Ave., 753-0000
Starker's, 200 Nichols Rd., 753-3565
Stephenson's Apple Farm, U.S. 40 & Lee's Summit Rd., 373-5400
Trader Vic's, in the Westin Crown Center, 391-4444

SELECTED ATTRACTIONS:
Harry S Truman Library & Museum, U.S. 24 & Delaware St., Independence, 833-1400
Truman Home National Historic Site, 223 N. Main St., Independence, 254-7199
Kansas City Museum, 3218 Gladstone Blvd., 483-8300
Kansas City Zoo, 6700 Zoo Dr. in Swope Park, 333-7408
Liberty Memorial, 100 W. 26th St., 221-1918
Nelson-Atkins Museum of Art, 4525 Oak St., 561-4000
Starlight Theater, 4600 Starlight Rd. in Swope Park, 363-7827
Watkins Woolen Mill State Historic Site, Rte. 2, Lawson, 296-3357
Worlds of Fun/Oceans of Fun (theme/water parks), 4545 Worlds of Fun Ave., 454-4545

INFORMATION SOURCES:
Convention and Visitors Bureau of Greater Kansas City
City Center Square
1100 Main St., Suite 2550
Kansas City, Missouri 64105
(816) 221-5242; (800) 767-7700
Visitor Info. Recording: (816) 691-3800
Chamber of Commerce of Greater Kansas City
920 Main St., Suite 600
Kansas City, Missouri 64105
(816) 221-2424

Las Vegas, Nevada

Population: (*741,459)
258,295 (1990C)
Altitude: 2,020 feet
Average Temp.: Jan., 44°F.; July, 90°F.
Telephone Area Code: 702
Time: 118   **Weather:** 734-2010
Time Zone: Pacific

AIRPORT TRANSPORTATION:
Eight miles to downtown Las Vegas.
Taxicab and limousine service.

SELECTED HOTELS:
Best Western Marianna Inn, 1322 E. Fremont St., 385-1150
Caesars Palace, 3570 Las Vegas Blvd. S., 731-7110

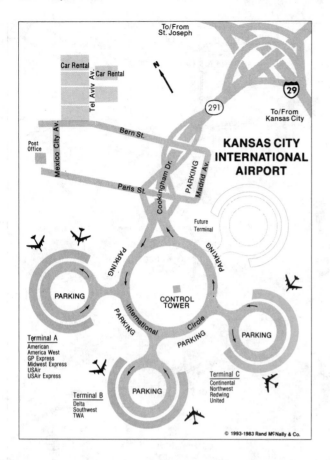

KANSAS CITY INTERNATIONAL AIRPORT

To/From St. Joseph

Car Rental
Car Rental

Post Office

Tel Aviv Av.
Mexico City Av.

Bern St.
Paris St.

Cookingham Dr.

PARKING
Madrid Av.

29
291

To/From Kansas City

Future Terminal

PARKING
PARKING
International Circle
PARKING

CONTROL TOWER

PARKING

PARKING

PARKING

Terminal A
American
America West
GP Express
Midwest Express
USAir
USAir Express

Terminal B
Delta
Southwest
TWA

Terminal C
Continental
Northwest
Redwing
United

© 1993-1983 Rand McNally & Co.

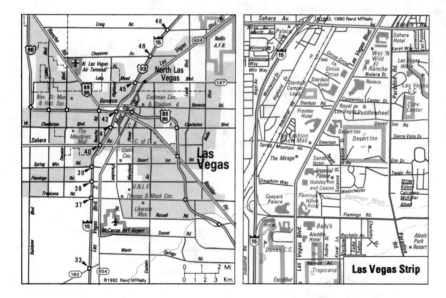

Las Vegas

Las Vegas Strip

©1992 Rand McNally

Tropicana, 3801 Las Vegas Blvd. S., 739-2222

Union Plaza Hotel/Casino, 1 Main St., 386-2110

SELECTED RESTAURANTS:
Bacchanal, in Caesars Palace Hotel, 734-7110
Georges La Forge's Pamplemousse Restaurant, 400 E. Sahara Ave., 733-2066
Golden Steer Steak House, 308 W. Sahara Ave., 384-4470
The House of Lords, in the Sahara Hotel, 737-2111
Nero's, in Caesars Palace Hotel, 731-7110
Palace Court, in Caesars Palace Hotel, 734-7110
Portofino Room, in the Desert Inn Hotel and Country Club, 733-4444
Regency Room, in the Sands Resort Hotel and Casino, 733-5000
The Rikshaw, in the Riviera Hotel, 2901 Las Vegas Blvd. S., 734-5110

SELECTED ATTRACTIONS:
Gambling and Entertainment, at casinos and hotels throughout Las Vegas
Hoover Dam, 30 miles east of Las Vegas on US 93, 293-1081
Liberace Museum, 1775 E. Tropicana Ave., 798-5595
Valley of Fire, 55 miles northeast of Las Vegas off I-15, 397-2088

INFORMATION SOURCES:
Las Vegas Convention and Visitors Authority
    Convention Center
    3150 Paradise Rd.
    Las Vegas, Nevada 89109
    (702) 892-0711
Las Vegas Chamber of Commerce
    711 E. Desert Inn
    Las Vegas, Nevada 89109
    (702) 735-1616

Los Angeles, California

Population: (*8,863,164) 3,485,398 (1990C)
Altitude: Sea level to 330 feet
Average Temp.: Jan., 55°F.; July, 73°F.
Telephone Area Code: 213
Time: 853-1212   **Weather:** 554-1212
Time Zone: Pacific

AIRPORT TRANSPORTATION:
Seventeen miles to downtown Los Angeles.
Taxicab, limousine bus and city bus service.

SELECTED HOTELS:
Beverly Hills Hotel, 9641 Sunset Blvd., (310) 276-2251
Biltmore Hotel, 506 S. Grand Ave., 624-1011
Century Plaza, 2025 Avenue of the Stars, (310) 277-2000
Holiday Inn—Hollywood, 1755 N. Highland Ave., 462-7181
Hotel Bel-Air, 701 Stone Canyon Rd., (310) 472-1211
Hyatt Los Angeles Airport, 6225 W. Century Blvd., (310) 670-9000

Desert Inn Hotel and Country Club, 3145 Las Vegas Blvd. S., 733-4444
Dunes Hotel & Country Club, 3650 Las Vegas Blvd. S., 737-4110
Excalibur, 3850 Las Vegas Blvd. S., 597-7777
Golden Nugget Hotel & Casino, 129 E. Fremont St., 385-7111
Hacienda Resort Hotel & Casino, 3950 Las Vegas Blvd. S., 739-8911

Imperial Palace Hotel & Casino, 3535 Las Vegas Blvd. S., 731-3311
Las Vegas Hilton, 3000 Paradise Rd., 732-5111
Maxim Hotel & Casino, 160 E. Flamingo, 731-4300
Sahara Hotel & Casino, 2535 Las Vegas Blvd. S., 737-2111
Sands Resort Hotel & Casino, 3355 Las Vegas Blvd. S., 733-5000

Los Angeles
California

scale

0 1 2 3 4 5 6 7 8 9 10 miles

© RAND McNALLY & CO.   PRINTED IN U.S.A.   92-1

Hyatt Regency Los Angeles, 711 S. Hope
   St., 683-1234
Kawada Hotel, 200 S. Hill St., 621-4455
L'Ermitage, 9291 Burton Way, 278-3344
Le Parc Hotel De Luxe, 733 N. West Knoll
   Dr., (310) 855-8888
Los Angeles Airport Marriott, 5855 W.
   Century Blvd., (310) 641-5700
Los Angeles Hilton and Towers, 930
   Wilshire Blvd., 629-4321
The New Otani Hotel & Garden, 120 S.
   Los Angeles St., 629-1200

Sheraton Townhouse Hotel, 2961
   Wilshire Blvd., 382-7171
Sheraton–Universal, 333 Universal
   Terrace Pkwy., Universal City,
   (818) 980-1212
University Hilton, 3540 S. Figueroa St.,
   748-4141
The Westin Bonaventure Hotel, 404 S.
   Figueroa St., 624-1000

SELECTED RESTAURANTS:
Bernard's, in the Biltmore Hotel, 612-1580

Epicentre Restaurant, in the Kawada
   Hotel, 621-4455
Lawry's Prime Rib, 55 N. La Cienega
   Blvd., (310) 652-2827
L'Escoffier, in the Beverly Hilton, 9876
   Wilshire Blvd., (310) 274-7777
L'Orangerie, 903 N. La Cienega Blvd.,
   (310) 652-9770
Madame Wu's Garden, 2201 Wilshire
   Blvd., (310) 828-5656
Pacific Dining Car, 1310 W. 6th St.,
   483-6000

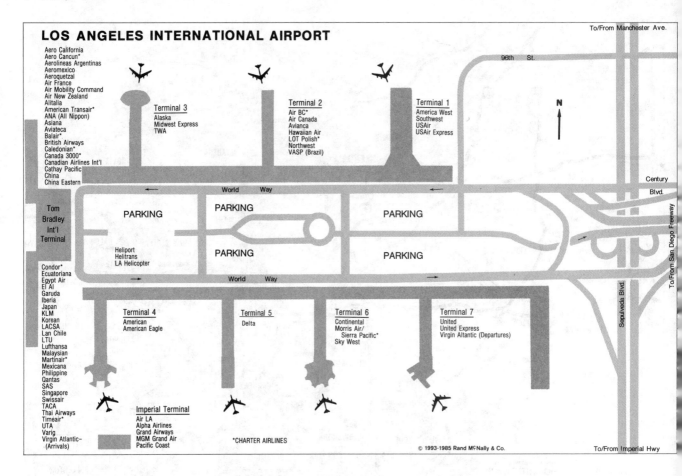

LOS ANGELES INTERNATIONAL AIRPORT

Aero California
Aero Cancun*
Aerolineas Argentinas
Aeromexico
Aeroquetzal
Air France
Air Mobility Command
Air New Zealand
Alitalia
American Transair*
ANA (All Nippon)
Asiana
Aviateca
Balair*
British Airways
Caledonian*
Canada 3000*
Canadian Airlines Int'l
Cathay Pacific
China
China Eastern

Terminal 3
Alaska
Midwest Express
TWA

Terminal 2
Air BC*
Air Canada
Avianca
Hawaiian Air
LOT Polish*
Northwest
VASP (Brazil)

Terminal 1
America West
Southwest
USAir
USAir Express

96th St.

N

Century Blvd.

Tom Bradley Int'l Terminal

PARKING

PARKING

PARKING

World Way

PARKING

PARKING

Heliport
Helitrans
LA Helicopter

World Way

To/From San Diego Freeway

Sepulveda Blvd.

Condor*
Ecuatoriana
Egypt Air
El Al
Garuda
Iberia
Japan
KLM
Korean
LACSA
Lan Chile
LTU
Lufthansa
Malaysian
Martinair*
Mexicana
Philippine
Qantas
SAS
Singapore
Swissair
TACA
Thai Airways
Timeair*
UTA
Varig
Virgin Atlantic–
(Arrivals)

Terminal 4
American
American Eagle

Terminal 5
Delta

Terminal 6
Continental
Morris Air/
    Sierra Pacific*
Sky West

Terminal 7
United
United Express
Virgin Atlantic (Departures)

Imperial Terminal
Air LA
Alpha Airlines
Grand Airways
MGM Grand Air
Pacific Coast

*CHARTER AIRLINES

© 1993-1985 Rand McNally & Co.

To/From Manchester Ave.

To/From Imperial Hwy

Stepps, 330 S. Hope St., 626-0900
Yamato, in the Century Plaza Hotel,
    (310) 277-1840

SELECTED ATTRACTIONS:
Disneyland, 1313 S. Harbor Blvd.,
    Anaheim, (714) 999-4000
El Pueblo de Los Angeles State Historic
    Park, 845 N. Alameda St., 628-1274
Farmer's Market & Shopping Village,
    6333 W. 3rd St., 933-9211
Griffith Park Observatory, 2800 E.
    Observatory Rd., 664-1191
Knott's Berry Farm, 8039 Beach Blvd.,
    Buena Park, (714) 220-5200
Los Angeles Zoo, 5333 Zoo Dr., 664-1100
Mann's Chinese Theatre (movie stars'
    hand and foot prints), 6925 Hollywood
    Blvd., 461-3331
Queen Mary/Spruce Goose, Pier J, Long
    Beach, (310) 435-3511
Rodeo Drive (shopping), Beverly Hills
Universal Studios, 100 Universal City
    Plaza, Universal City, (818) 777-3750

INFORMATION SOURCES:
Los Angeles Convention & Visitors
Bureau
    515 S. Figueroa, 11th Floor
    Los Angeles, California 90071
    (213) 624-7300
Los Angeles Chamber of Commerce
    404 S. Bixel St.
    Los Angeles, California 90017
    (213) 629-0602

Louisville, Kentucky

Population: (*952,662)
    269,063 (1990C)
Altitude: 462 feet
Average Temp.: Jan., 35°F.; July, 78°F.
Telephone Area Code: 502
Time: 585-5961 **Weather:** 363-9655
Time Zone: Eastern

AIRPORT TRANSPORTATION:
Five miles to downtown Louisville.
Taxicab, limousine, and bus service.

SELECTED HOTELS:
Breckinridge Inn, 2800 Breckinridge Lane,
    456-5050
The Brown Hotel, 335 W. Broadway,
    583-1234
Executive Inn, 978 Phillips Ln., 367-6161
Executive West Hotel, 830 Phillips Ln.,
    367-2251
The Galt House, 140 N. 4th Ave.,
    589-5200
Holiday Inn Louisville/Downtown, 120 W.
    Broadway, 582-2241
Holiday Inn—S.W., 4110 Dixie Hwy.,
    448-2020
Hurstbourne Hotel & Conference Center,
    9700 Bluegrass Pkwy., 491-4830
Hyatt Regency Louisville, 320 W.
    Jefferson St., 587-3434
Quality Hotel, 100 E. Jefferson St.,
    582-2481
The Seelbach Hotel, 500 Fourth Ave.,
    585-3200

SELECTED RESTAURANTS:
The Atrium, 1028 Barret Ave., 456-6789
Casa Grisanti, 1000 E. Liberty St.,
    584-4377
Mama Grisanti, 3938 DuPont Circle,
    893-0141
New Orleans House, 9424 Shelbyville
    Rd., 426-1577
Oak Room, in The Seelbach Hotel,
    585-3200
The Spire, in the Hyatt Regency
    Louisville, 587-3434

SELECTED ATTRACTIONS:
Actors Theatre of Louisville, 316 W. Main
    St., 584-1205
Churchill Downs, 700 Central Ave.,
    636-4400
Farmington Historic Home, 3033
    Bardstown Rd., 452-9920
Hillerich & Bradsby (Louisville Slugger
    factory), Slugger Park, 1525
    Charlestown-New Albany Rd.,
    Jeffersonville, Indiana, 585-5226
J.B. Speed Art Museum, 2035 S. Third
    St., 636-2893
Kentucky Center for the Arts, W. Main St.
    between 5th & 6th, 584-7777
Kentucky Derby Museum, 704 Central
    Ave., 637-1111
Locust Grove Historic Home, 561
    Blankenbaker Ln., 897-9845
Louisville Zoo, 1100 Trevilian Way,
    459-2181
Museum of History & Science/IMAX
    Theater, 727 W. Main St., 561-6100

INFORMATION SOURCES:
Louisville Convention & Visitors Bureau
400 S. 1st
Louisville, Kentucky 40202
(502) 584-2121
Louisville Tourist Information
400 S. 1st
Louisville, Kentucky 40202
(502) 582-3732
Louisville Area Chamber of Commerce
1 Riverfront Plaza
Louisville, Kentucky 40202
(502) 566-5000

Memphis, Tennessee

Population: (*981,747)
610,337 (1990C)
Altitude: 264 feet
Average Temp.: Jan., 43°F.; July, 80°F.
Telephone Area Code: 901
Time: 526-5261 **Weather:** 756-4141
Time Zone: Central

AIRPORT TRANSPORTATION:

Ten miles to downtown Memphis.
Taxicab and limousine bus service.

SELECTED HOTELS:

Brownestone Hotel, 300 N. Second St.,
525-2511
Holiday Inn Crowne Plaza, 250 N. Main,
527-7300
Holiday Inn Memphis Airport, 1441 E.
Brooks Rd., 398-9211
Holiday Inn Overton Square, 1837 Union
Ave., 278-4100
Memphis Airport Hotel, 2240 Democrát
Rd., 332-1130
Memphis Marriott, 2625 Thousand Oaks
Blvd., 362-6200
Omni Hotel, 939 Ridge Lake Blvd.,
684-6664
The Peabody Hotel, 149 Union Ave.,
529-4000
Sheraton Memphis Airport, at the
Airport, 332-2370

SELECTED RESTAURANTS:

Benihana of Tokyo, 912 Ridgelake,
683-7390
Captain Bilbo's River Restaurant, 263
Wagner Pl., 526-1966
Grisanti's, 1489 Airways Blvd., 458-2648
Justine's, 919 Coward Pl., 527-3815
Paulette's, 2110 Madison Ave., 726-5128
The Pier, 100 Wagner Pl., 526-7381
The Rendezvous, 52 S. 2nd St., 523-2746

SELECTED ATTRACTIONS:

Beale Street (birthplace of the blues),
526-0110
Graceland, 3764 Elvis Presley Blvd.,
332-3322
The Great American Pyramid, Front &
Auction, 529-1991
Liberty Land/Liberty Bowl Complex, 940
Early Maxwell Blvd., 274-1776
Memphis International Sports Park, 5500
Taylor Forge Rd., Millington, 358-7223
Memphis Pink Palace Museum &
Planetarium, 3050 Central Ave.,
320-6320
Memphis Queen Line, 527-5694
Memphis Zoo & Aquarium, 2000
Gallaway, 726-4775
National Civil Rights Museum, 450
Mulberry, 521-9699

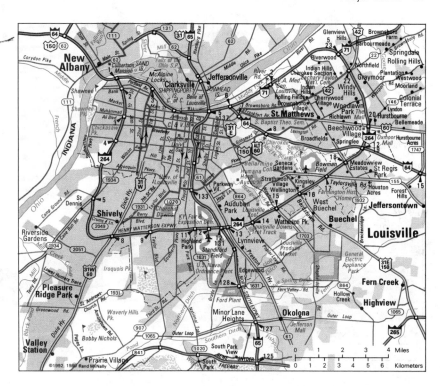

INFORMATION SOURCES:
Memphis Convention & Visitors Bureau
50 N. Front, Suite 450
Memphis, Tennessee 38103
(901) 576-8181
Memphis Area Chamber of Commerce
22 N. Front, Suite 200 Falls Building
Memphis, Tennessee 38103
(901) 575-3500

Mexico City (Ciudad de México), Mexico

Population: (*18,000,000)
9,377,300 (1991E)
Altitude: 7,450 ft.
Average Temp.: Jan., 54°F.; July, 64°F.
Telephone Code: 011-52-5
Time Zone: Central

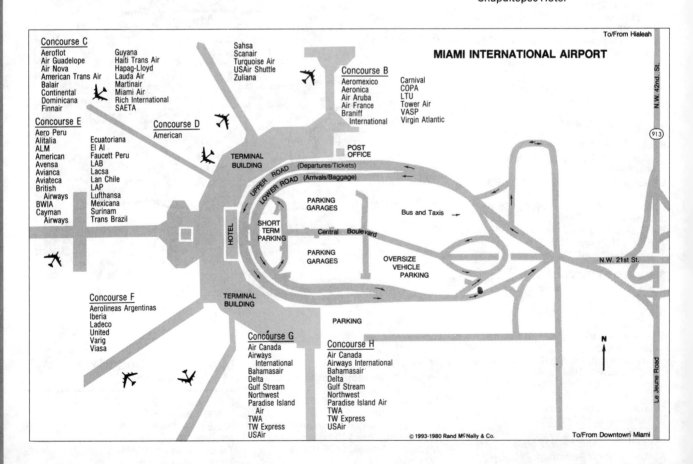

AIRPORT TRANSPORTATION:
Four miles to downtown Mexico City. Taxicab and limousine service.

SELECTED HOTELS:
Aristos, Paseo de la Reforma 276
Camino Real, Av. Mariano Escobedo 700
Crown Plaza, Paseo de la Reforma 80
Del Prado Hotel, Marina Nacional 399
Galeria Plaza, Hamburgo 159
Gran Hotel Howard Johnson, 16 de Septiembre No. 82, Zocalo
Hotel Century, Liverpool 152
Hotel Nikko, Campos Eliseos 204
Krystal Zona Rosa, Liverpool 155
Maria Isabel-Sheraton, Paseo de la Reforma 325
Radisson Paraiso, Cúspide 53
Stouffer Presidente, Campos Eliseos 218

SELECTED RESTAURANTS:
Anderson's, Paseo de la Reforma 382
Chalet Suizo, Niza 37
Champs Elysses, Amberes 1 Piso 1
El Parador, Niza 17
Focolare, Hamburgo 87 Zona Rosa
Fonda Santa Anita, Londres 38
Fouquets de Paris, in the Camino Real Hotel
Hacienda de Tlalpan, Calzada de Tlalpan 4619
Hacienda de los Morales, Vázquez de Mella 525
La Cava, Insurgentes Sur 2465
Loredo, Hamburgo 29
Maxims, in the Hotel El Presidente Chapultepec Hotel

Mesón del Caballo Bayo, Av. del
 Conscripto 360
Restaurant del Lago, Chapultepec Park
 2nd Section
Rivoli, Hamburgo 123
San Angel Inn, Diego Rivera 50, San
 Angel

SELECTED ATTRACTIONS:

Ballet Folklórico, at the Palace of Fine
 Arts on Ave. Juárez
Chapultepec Castle, at the entrance to
 Chapultepec Park
Floating Gardens of Xochimilco
Hipodromo de las Americas
 (horseracing)
National Museum of Anthropology, in
 Chapultepec Park
National Palace
Polyforum Cultural Siqueiros,
 Insurgentes, corner of Filadelfis
Pyramids of Teotihuacan (light and
 sound show)
Rufino Tamayo Museum (art), in
 Chapultepec Park on the west side of
 Reforma
Templo Mayor (Aztec temple remains),
 northeast corner of the Zocalo

INFORMATION SOURCES:

Ministry of Tourism
 Mariano Escobedo 726
 Colonia Anzures
 Del. Miguel Hidalgo 11590
 México, D.F.
Mexican Government Tourist Office
 405 Park Avenue, Suite 1401
 New York, New York 10022
 (212) 755-7261; (800) 262-8900

Miami, Florida

Population: (*1,937,094)
 358,548 (1990C)
Altitude: Sea level to 30 feet
Average Temp.: Jan., 69°F.; July, 82°F.
Telephone Area Code: 305
Time: 324-8811   **Weather:** 661-5065
Time Zone: Eastern

AIRPORT TRANSPORTATION:

Five miles to downtown Miami.
Taxicab and limousine bus service.

SELECTED HOTELS:

Best Western Marina Park, 340 Biscayne
 Blvd., 371-4400
Biscayne Bay Marriott, 1633 N. Bayshore
 Dr., 374-3900
The Colonnade, 180 Aragon Ave., Coral
 Gables, 441-2600
Don Shula's Hotel & Golf Club, Main St.
 & Bull Run Rd., Miami Lakes, 821-1150
Doubletree, 2649 S. Bayshore Dr.,
 858-2500
Fontaine Bleau Hilton, 4441 Collins Ave.,
 Miami Beach, 538-2000
Holiday Inn—Civic Center, 1170 NW. 11th
 St., 324-0800
Hyatt Regency Miami, 400 SE. 2nd Ave.,
 358-1234
Marriott Airport Hotel, 1201 NW. LeJeune
 Rd., 649-5000
Miami Airport Hilton, 5101 Blue Lagoon
 Dr., 262-1000

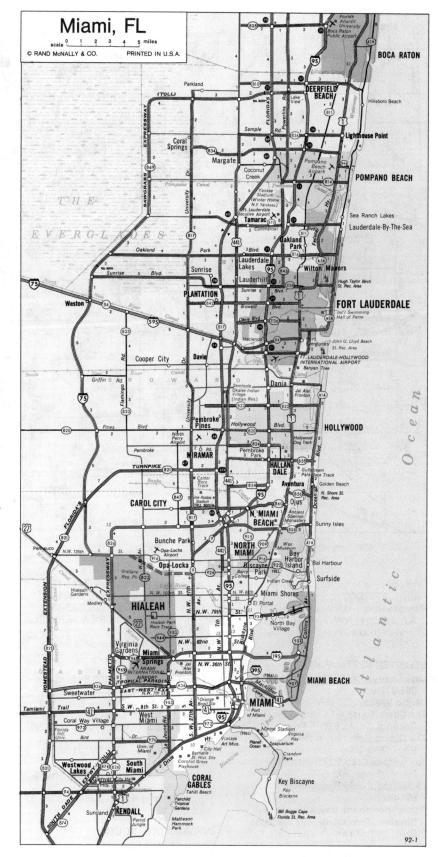

Omni International Hotel, 1601 Biscayne
Blvd., 374-0000
Ramada Hotel—Miami International
Airport, 3941 NW. 22nd St., 871-1700
Sheraton Brickell Point, 495 Brickell Ave.,
373-6000

SELECTED RESTAURANTS:

Cafe Chauveron, 9561 E. Bay Harbor Dr.,
866-8779
Centro Vasco, 2235 SW. 8th St., 643-9606
The Chart House, 51 Charthouse Dr.,
Coconut Grove, 856-9741
The Dining Galleries, in the Fontaine
Bleau, 538-2000
Joe's Stone Crab, 227 Biscayne St., South
Miami Beach, 673-0365
La Paloma, 10999 Biscayne Blvd.,
891-0505
Reflections on the Bay, at the back of
Bayside Marketplace, 401 Biscayne
Blvd., 371-6433

SELECTED ATTRACTIONS:

Ancient Spanish Monastery, 16711 W.
Dixie Hwy., North Miami Beach,
945-1461
Biscayne National Park Tour Boats,
Biscayne National Park Headquarters, 9

mi. east of Homestead, 247-2400
Fairchild Tropical Garden, 10901 Old
Cutler Rd., Coral Gables, 667-1651
Miami Metrozoo, 12400 SW 152nd St.,
251-0400
Miami Museum of Science & Space
Transit Planetarium, 3280 S. Miami
Ave., 854-4247
Miami Seaquarium, 4400 Rickenbacker
Causeway, 361-5705
Miccosukee Indian Village & Airboat
Tours, 30 miles west of Miami on U.S.
41, 223-8380 (weekdays) or 223-8388
(weekends)
Monkey Jungle, 14805 SW 216th St.,
235-1611
Parrot Jungle & Gardens, 11000 SW 57th
Ave., 666-7834
Vizcaya Museum and Gardens, 3251 S.
Miami Ave., 579-2813

INFORMATION SOURCE:

Greater Miami Convention and Visitors
Bureau
701 Brickell Ave., Suite 2700
Miami, Florida 33131
(305) 539-3000; (800) 933-8448

Milwaukee, Wisconsin

Population: (*1,432,149)
628,088 (1990C)
Altitude: 634 feet
Average Temp.: Jan., 21°F.; July, 71°F.
Telephone Area Code: 414
Time: 844-0443 **Weather:** 936-1212
Time Zone: Central

AIRPORT TRANSPORTATION:

Eight miles to downtown Milwaukee.
Taxicab, bus, and limousine bus service.

SELECTED HOTELS:

The Grand Milwaukee Hotel, 4747 S.
Howell Ave., 481-8000
Hyatt Regency Milwaukee, 333 W.
Kilbourn Ave., 276-1234
Marc Plaza Hotel, 509 W. Wisconsin Ave.,
271-7250
Milwaukee River Hilton Inn, 4700 N. Port
Washington Rd., 962-6040
The Pfister Hotel, 424 E. Wisconsin Ave.,
273-8222
Sheraton Mayfair Inn, 2303 N. Mayfair,
Wauwatosa, 257-3400
Wyndham Hotel, 139 E. Kilbourn Ave.,
276-8686

SELECTED RESTAURANTS:

Benson's, in the Marc Plaza Hotel,
271-7250
The English Room, in The Pfister Hotel,
273-8222
Grenadier's, 747 N. Broadway, 276-0747
John Ernst's, Ogden at Jackson, 273-1878
Karl Ratzsch's, 320 E. Mason St.,
276-2720
Mader's German Restaurant, 1037 N. 3rd
St., 271-3377
Pieces of Eight, 550 N. Harbor Dr.,
271-0597
Whitney's, in the Milwaukee Marriott, 375
S. Moorland Rd., Brookfield, 786-1100

SELECTED ATTRACTIONS:

Miller Brewing Company, 4251 W. State
St., 931-BEER
Milwaukee Art Museum, 750 N. Lincoln
Memorial Dr., 224-3200
Milwaukee County Zoo, 10001 W.
Bluemound Rd., 771-3040
Milwaukee Public Museum, 800 W. Wells
St., 278-2702
Mitchell Park Horticultural Conservatory
(The Domes), 524 S. Layton Blvd.,
649-9830
Pabst Brewing Company, 915 W. Juneau
Ave., 223-3709
Pabst Mansion, 2000 W. Wisconsin Ave.,
931-0808

INFORMATION SOURCES:

Greater Milwaukee Convention & Visitors
Bureau
510 W. Kilbourn Ave.
Milwaukee, Wisconsin 53203
(414) 273-3950; (800) 231-0903
Metropolitan Milwaukee Association of
Commerce
756 N. Milwaukee St.
Milwaukee, Wisconsin 53202
(414) 273-3000

Minneapolis-St. Paul, Minnesota

Population: (*2,464,124)
(Minneapolis) 368,383;
(St. Paul) 272,235 (1990C)
Altitude: (Minneapolis) 840 feet;
(St. Paul) 874 feet
Average Temp.: Jan., 15°F.; July, 74°F.
Telephone Area Code: 612
Time: 546-8463 **Weather:** 452-2323
Time Zone: Central

AIRPORT TRANSPORTATION:

About 8 miles to downtown Minneapolis
or St. Paul.
Taxicab, bus, and limousine bus service
to Minneapolis and St. Paul.

SELECTED HOTELS: MINNEAPOLIS

Bloomington Marriott Hotel, 2020 E. 79th
St., 854-7441
Holiday Inn—Airport #2, 5401 Green
Valley Dr., 831-8000
Hotel Sofitel, 5601 W. 78th St., 835-1900
Hyatt Regency Minneapolis, 1300
Nicollet Mall, 370-1234
Minneapolis Marriott City Center, 30 S.
7th St., 349-4000
Omni Northstar, 618 2nd Ave. S,
338-2288

Radisson Hotel South, 7800 Normandale
Blvd., 835-7800
Registry, 7901 24th Ave. S., 854-2244
Sheraton Airport Inn, 2500 E. 79th St.,
854-1771
The Vista Marquette, 7th St. & Marquette
Ave., 332-2351
Wyndham Garden Hotel, 4460 W. 78th St.
Circle, 831-3131

SELECTED RESTAURANTS: MINNEAPOLIS

The Anchorage Restaurant, 1330
Industrial Blvd., 379-4444
Faegre's, 430 1st Ave. N., 332-3515
Goodfellow's, 800 Nicollet Mall, 332-4800
Gustino's, in the Minneapolis Marriott
City Center, 349-4000
Lord Fletcher's of the Lake, 3746 Sunset
Dr., Spring Park, 471-8513
Taxi Restaurant, in the Hyatt Regency
Minneapolis, 370-1234

SELECTED ATTRACTIONS: MINNEAPOLIS

American Swedish Institute, 2600 Park
Ave., 871-4907
Bell Museum (Minnesota flora and fauna
in natural habitat), University of
Minnesota, 10 Church St. SE, 624-1852
Guthrie Theatre, 725 Vineland Pl.,
377-2224
The Mall of America, MN Hwy. 77 & I-494,
Bloomington, 851-3500
Minneapolis Institute of Arts, 2400 Third
Ave. S., 870-3046
Minneapolis Planetarium, 300 Nicollet
Mall, 372-6644
Nicollet Mall, downtown Minneapolis
Riverplace (European-theme festival
marketplace), 1 Main St. SE, 378-1969

INFORMATION SOURCES: MINNEAPOLIS

Greater Minneapolis Convention &
Visitor Association
1219 Marquette
Minneapolis, Minnesota 55403
(900) 860-0092 (tourist information:
$1.99 first minute; 99¢ each additional
minute)

SELECTED HOTELS: ST. PAUL

Best Western Kelly Inn State Capitol, 161
St. Anthony Blvd., 227-8711
Crown Sterling Suites, 175 E. 10th St.,
224-5400
Holiday Inn St. Paul East, I-94 at
McKnight, 731-2220
The Radisson Hotel St. Paul, 11 E.
Kellogg Blvd., 292-1900
Ramada Hotel St. Paul, 1870 Old Hudson
Rd., 735-2330
The St. Paul Hotel, 350 Market St.,
292-9292
Sheraton Midway, 400 N. Hamline Ave.,
642-1234
Sunwood Inn at Bandana Square, 1010
Bandana Blvd., 647-1637

SELECTED RESTAURANTS: ST. PAUL

Bali Hai, 2305 White Bear Ave. and Hwy
36, Maplewood, 777-5500
Forepaugh's Restaurant, 276 S. Exchange
St., 224-5606
Gallivan's, 354 Wabasha St., 227-6688
Le Carrousel, in The Radisson Hotel St.
Paul, 292-1900

McGovern's Pub, 225 W. 7th St., 224-5821
Venetian Inn, 2814 Rice St., 484-7215

SELECTED ATTRACTIONS: ST. PAUL

Alexander Ramsey House, 265 S.
Exchange St., 296-8681
Children's Museum, 1217 Bandana Blvd.
N., 644-5305
Como Park Zoo and Conservatory,
Midway Pkwy. & Kaufman Dr.,
488-5571
Historic Fort Snelling, Fort Rd. at MN
Hwys. 5 & 55, 726-9430
James J. Hill House, 240 Summit Ave.,
297-2555
Jonathan Padelford & Josiah Snelling
Riverboats, Harriet Island, 227-1100
Minnesota History Center, 690 Cedar St.,
297-1827
Minnesota Museum of Art, 305 Saint
Peter St., 292-4355
Minnesota State Capitol, Cedar & Aurora
sts., 297-3521 (recorded info.);
296-2881 (tour reservations)
Science Museum of Minnesota/William L.
McKnight Omnitheater, 30 E. 10th St.,
221-9454

INFORMATION SOURCES: ST. PAUL

St. Paul Convention & Visitors Bureau
101 Norwest Center
55 E. 5th St.
St. Paul, Minnesota 55101
(612) 297-6985; (800) 627-6101
Minnesota Office of Tourism
375 Jackson St., 250 Skyway Level
St. Paul, Minnesota 55101
(612) 296-5029; (800) 657-3700
St. Paul Area Chamber of Commerce
101 Norwest Center
55 E. 5th St.
St. Paul, Minnesota 55101
(612) 223-5000

Montréal, Québec, Canada

Population: (*2,921,357)
1,015,420 (1986C)
Altitude: 50 ft.
Average Temp.: Jan., 16°F.; July, 71°F.
Telephone Area Code: 514
Time: (none) **Weather:** 636-3026
Time Zone: Eastern

AIRPORT TRANSPORTATION:

Fourteen miles from Dorval Airport to
downtown Montréal, 34 miles from
Mirabel Airport.
Taxicab and bus service.
Transit service between Dorval and
Mirabel airports.

SELECTED HOTELS:

Aeroport Hilton International, 12505 Cote
de Liesse Rd., 631-2411
Bonaventure Hilton International,
Mansfield & Lagauchetière, 878-2332
Centre Sheraton, 1201 René-Lévesque
Blvd. W., 878-2000
Four Seasons Hotel, 1050 Sherbrooke St.
W., 284-1110
Grand Hotel, 777 University St., 879-1370
Hotel Meridien-Montréal, 4 Complexe
Desjardins, 285-1450
Le Chateau Champlain, 1050 Ouest de
Lagauchetière, 878-9000

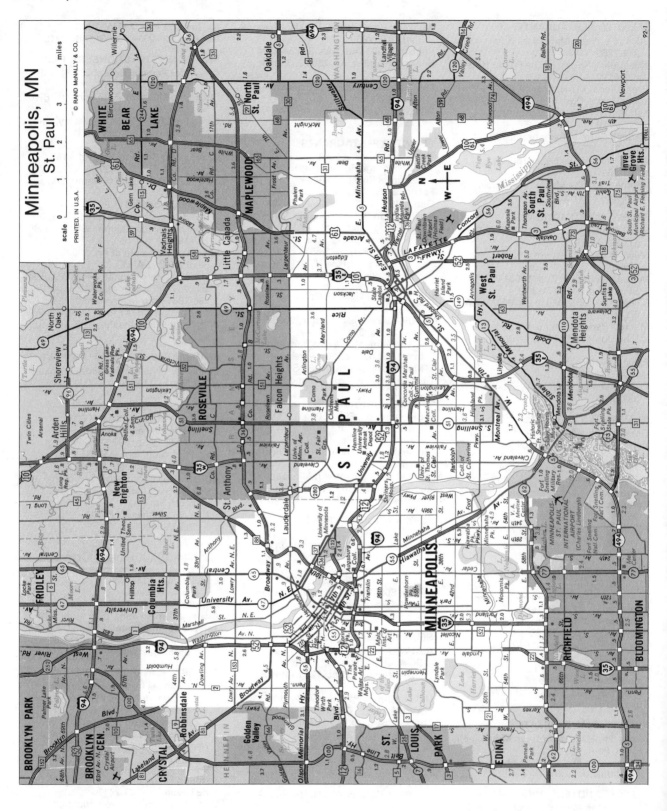

Minneapolis, MN
St. Paul

© RAND McNALLY & CO.

scale   0   1   2   3   4 miles

PRINTED IN U.S.A.

Queen Elizabeth, 900 Blvd.
  René-Lévesque, 861-3511
Ritz-Carlton, 1228 Sherbrooke St. W.,
  842-4212
Ruby Foo's Hotel, 7655 Decarie Blvd.,
  731-7701

SELECTED RESTAURANTS:

Beaver Club, in the Queen Elizabeth
  Hotel, 861-3511
Café de Paris, in the Ritz-Carlton,
  842-4212
Chez la Mère Michel, 1209 Guy, 934-0473
Desjardins, 1175 Mackay, 866-9741
Le Castillon, in the Bonaventure Hilton
  International, 878-2332
Le Neufchatel, in Le Chateau Champlain,
  878-9000
Les Filles du Roy, 415 Bonsecours,
  849-3535
Les Halles, 1450 Crescent, 844-2328

SELECTED ATTRACTIONS:

Botanical Gardens, 4101 Sherbrooke St.
  E., 872-1427
Canadian Centre for Architecture, 1920
  Baile St., 939-7000
Insectarium, 4101 Sherbrooke St. E.,
  872-0663
Montréal Museum of Fine Arts, 1379
  Sherbrooke St. W., 285-1600
Notre Dame Basilica, in front of Place
  d'Armes, Old Montréal, 842-2925
Olympic Park and Tower, 3200 Viau,
  252-8687
St. Joseph's Oratory, 3800 Queen Mary
  Rd., 733-8211

INFORMATION SOURCES:

Greater Montréal Convention & Tourism
Bureau,
  1555 Peel St., Suite 600
  Montréal, Québec H3A 1X6
  (514) 844-5400
Canadian Consulate General
  1251 Avenue of the Americas
  New York, New York 10020-1175
  (212) 768-2400

Nashville, Tennessee

Population: (*985,026)
  487,969 (1990C)
Altitude: 440 feet
Average Temp.: Jan., 38°F.; July, 80°F.
Telephone Area Code: 615
Time: 259-2222 **Weather:** 244-9393
Time Zone: Central

AIRPORT TRANSPORTATION:

Seven miles to downtown Nashville.
Taxicab, limousine and bus service.

SELECTED HOTELS:

Airport Quality Inn, 1 International Plaza,
  361-7666
Doubletree Hotel, 315 4th Ave. N.,
  244-8200
The Hermitage, 231 6th Ave. N., 244-3121
Holiday Inn—Briley Pkwy., 2200 Elm Hill
  Pike, 883-9770
Holiday Inn Express—Southeast Airport,
  981 Murfreesboro Rd., 367-9150
Nashville Airport Marriott Hotel, One
  Marriott Dr., 889-9300
Opryland Hotel, 2800 Opryland Dr.,
  889-1000

Quality Inn Hall of Fame, 1407 Division
  St., 242-1631
Ramada Inn Downtown, 840 James
  Robertson Pkwy., 244-6130
Ramada Inn—Southwest Airport, 709
  Spence Ln., 361-0102
Ramada South Inn & Convention Center,
  737 Harding Pl., 834-5000
Regal Maxwell House, 2025 MetroCenter
  Blvd., 259-4343
Stouffer Nashville, 611 Commerce St.,
  255-8400

SELECTED RESTAURANTS:

Arthur's of Nashville, in the Union
  Station Hotel, 1001 Broadway,
  255-1494
Crown Court, in the Regal Maxwell
  House, 259-4343

The Hermitage Hotel Dining Room, in
  The Hermitage Hotel, 244-3121
Hunt Room, in the Doubletree Hotel,
  244-8200
Julian's Restaurant Francais, 2412 West
  End Ave., 327-2412
Mario's, 2005 Broadway, 327-3232
New Orleans Manor, 1400 Murfreesboro
  Rd., 367-2777
Old Hickory, in the Opryland Hotel,
  889-1000
Praline's, in the Regal Maxwell House,
  259-4343
Stockyard, 901 2nd Ave. N., 255-6464

SELECTED ATTRACTIONS:

Belle Meade Mansion, 110 Leake Ave.,
  356-0501
Cheekwood, Forest Park Dr., 356-8000

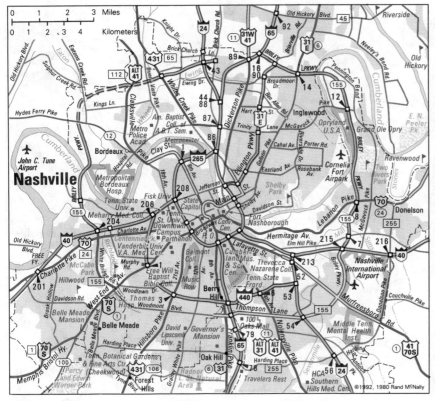

Country Music Hall of Fame, 4 Music
Square E., 256-1639
Cumberland Science Museum, 800 Ridley
Blvd., 862-5160
General Jackson Showboat, exit 11 on
Briley Pkwy., 889-6611
Grand Ole Opry, 2808 Opryland Dr.,
889-6600
The Hermitage, 4580 Rachel's Lane,
Hermitage, 889-2941
Opryland USA, 2802 Opryland Dr.,
889-6600
The Parthenon, West End & 25th aves.,
862-8431
Tennessee State Museum, Polk Cultural
Center, 505 Deaderick St., 741-2692

INFORMATION SOURCES:

Nashville Convention & Visitors Bureau
161 4th Ave. N.
Nashville, Tennessee 37219
(615) 259-4730
Nashville Area Chamber of Commerce
161 4th Ave. N.
Nashville, Tennessee 37219
(615) 259-4755

New Orleans, Louisiana

Population: (*1,238,816)
496,938 (1990C)
Altitude: [accent]5 to 25 feet
Average Temp.: Jan., 55°F.; July, 82°F.
Telephone Area Code: 504
Time: 976-1111   **Weather:** 465-9212
Time Zone: Central

AIRPORT TRANSPORTATION:

Eleven miles to downtown New Orleans.
Taxicab, airport shuttle, and limousine
bus service.

SELECTED HOTELS:

Clarion Hotel, 1500 Canal St., 522-4500
Dauphine Orleans, 415 Dauphine St.,
586-1800
Fairmont Hotel, 123 Baronne St.,
529-7111
Hotel Intercontinental, 444 St. Charles
Ave., 525-5566
Hyatt Regency New Orleans, 500 Poydras
Plaza, 561-1234
The Monteleone, 214 Royal St., 523-3341
New Orleans Hilton Riverside and
Towers, Poydras St. at the Mississippi
River, 561-0500
New Orleans Marriott, 555 Canal St.,
581-1000
The Omni Royal Orleans Hotel, 621 St.
Louis St., 529-5333
The Pontchartrain Hotel, 2031 St. Charles
Ave., 524-0581
Royal Sonesta Hotel, 300 Bourbon St.,
586-0300

SELECTED RESTAURANTS:

Arnaud's, 813 Rue Bienville, 523-5433
Brennan's, 417 Royal St., 525-9711
Broussard's, 819 Conti St., 581-3866
Caribbean Room, in The Pontchartrain
Hotel, 524-0581
Commander's Palace Restaurant, 1403
Washington Ave., 899-8221
Copeland's, 1001 S. Clearview Pkwy.,
733-7843
Galatoire's Restaurant, 209 Bourbon St.,
525-2021

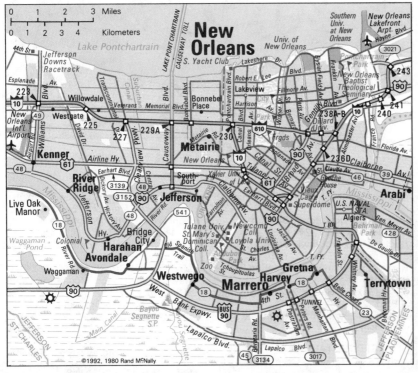

Kabby's, in the New Orleans Hilton
 Riverside and Towers, 561-0500
Louis XVI French Restaurant, in the St.
 Louis Hotel, 730 Bienville, 581-7000
Rib Room, in The Omni Royal Orleans
 Hotel, 529-5333
Sazerac Restaurant, in the Fairmont
 Hotel, 529-7111

SELECTED ATTRACTIONS:

Aquarium of the Americas, 1 Canal St.,
 565-3033
Audubon Zoo, 6500 Magazine St.,
 861-2537
French Quarter, 78-block area bounded
 by the Mississippi River, Esplanade
 Ave., Rampart St., & Canal St.
Jax Brewery, 620 Decatur St., 586-8015
New Orleans Museum of Art, Lelong Ave.
 in City Park, 488-2631
New Orleans Paddlewheels (cruises),
 529-4567
New Orleans Steamboat Company
 (cruises), 586-8777
Riverwalk (festival marketplace), 1
 Poydras St., 522-1555
New Orleans Centre, 1400 Poydras St.,
 568-0000

INFORMATION SOURCES:

Greater New Orleans Tourist &
Convention Commission
 1520 Sugar Bowl Dr.
 New Orleans, Louisiana 70112
 (504) 566-5011
The Chamber/New Orleans and the River
Region
 301 Camp St.
 New Orleans, Louisiana 70130
 (504) 527-6900

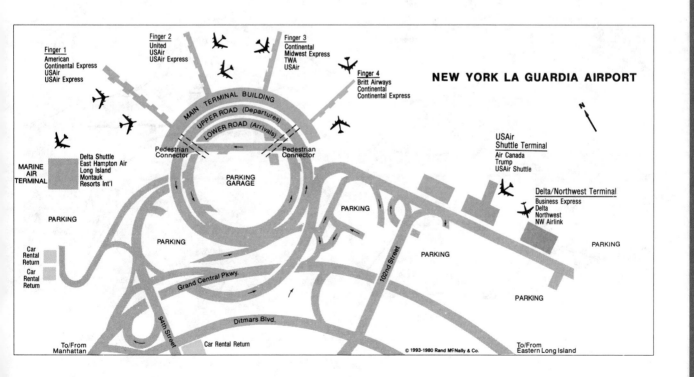

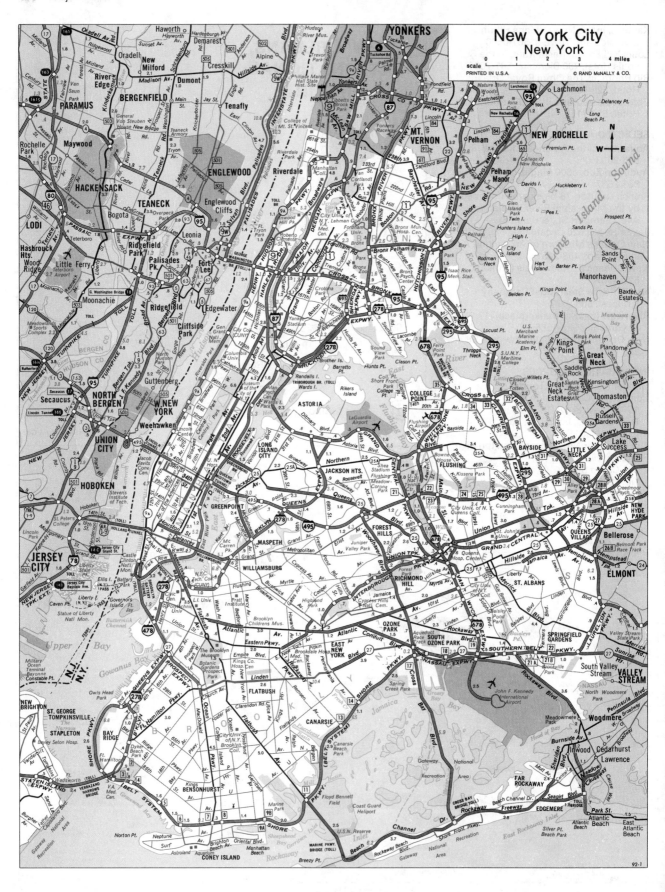

New York City
New York

NEW YORK JOHN F. KENNEDY INTERNATIONAL AIRPORT

N

Car Rental

150th Street

Car Rental

American Terminal
American
Finn Air
Qantas
Sabena

British Airways Terminal
British Airways
United

Trans World Terminal (Domestic Flights)
Jet/Trans World Express
Philippine
TWA(Domestic)

To/From Southern Parkway, Long Term Parking And Downtown New York City

Terminal Three
American
Eagle
Northwest

Trans World Terminal (International Flights)
Avianca
China·
Ecuatoriana
NY Helicopter
Transbrasil
TWA(Int'l)
USAir
USAir Express

Van Wyck Expressway

PARKING

PARKING

Car Rental

PARKING

PARKING

General Aviation Terminal Bldg.

CONTROL TOWER

Terminal One
America West
American Trans Air
Avensa
Baltia
Braniff
Key
Ladeco
Miami Air
MGM Grand
North American
Rich
Surinam
Tower Air

PARKING

INTERNATIONAL TERMINAL

East Wing
Aer Lingus
Aerolineas Argentinas
Aeromexico
Air Europa
Air Jamaica
Austrian
BWIA
Condor
Guyana
Iberia
Icelandair
Japan
KLM
Korean
Kuwait

Lacsa
LTU
Lufthansa
Martinair
Mexicana
Olympic
Royal Air Maroc
South African
TAP
Turkish
Varig
Virgin Atlantic

Terminal One-A
Carnival
Czechoslovak
Delta
Saudi
TAROM-Romanian

Delta Terminal
ANA
Aeroflot
CAAC
Delta
Jes Air
Malev

West Wing
Aero Cancun
Air Afrique
Air America
Air France
Air India
ALIA-Royal
Jordanian
Alitalia
Austrian
BWIA
Cayman
Dominica
Egypt Air
EL AI
Latur

LOT Polish
Nigeria
Pakistan
SAS
Spanair
Swissair
TACA
Taesa
Trans Continental
Viasa
Yugoslav

ROOFTOP PARKING

© 1993-1983 Rand McNally & Co.

New York, New York

Population: (*8,546,846)
7,322,564 (1990C)
Altitude: Sea level to 30 feet
Average Temp.: Jan., 33°F.; July, 75°F.
Telephone Area Code: 212
Time: 976-1616 **Weather:** 976-1212
Time Zone: Eastern

AIRPORT TRANSPORTATION:

Fifteen miles to Manhattan from JFK Airport; 8 miles from La Guardia Airport to Manhattan; 10 miles from Newark Airport to Manhattan.

Taxicab; limousine bus service to and from JFK, La Guardia, and Newark airports and East Side Airlines Terminal. Also JFK Express Subway/bus service from Manhattan to JFK Airport.

SELECTED HOTELS:

Carlyle, Madison Ave. at E. 76th St., 744-1600
The Helmsley Palace, 455 Madison Ave., 888-7000
The Hotel Pierre, 2 E. 61st St. at 5th Ave., 838-8000

Marriott Marquis, 1535 Broadway at 46th St., 398-1900
The New York Hilton & Towers at Rockefeller Center, 1335 Avenue of the Americas, 586-7000
The Plaza, 768 5th Ave., 759-3000
Regency, 540 Park Ave., 759-4100
Sheraton New York, 811 7th Ave. at 52nd St., 581-1000
Sherry-Netherland, 781 5th Ave., 355-2800
United Nations Plaza Park Hyatt, 1 U.N. Plaza, 355-3400
Waldorf-Astoria, 301 Park Ave. at 50th St., 355-3000

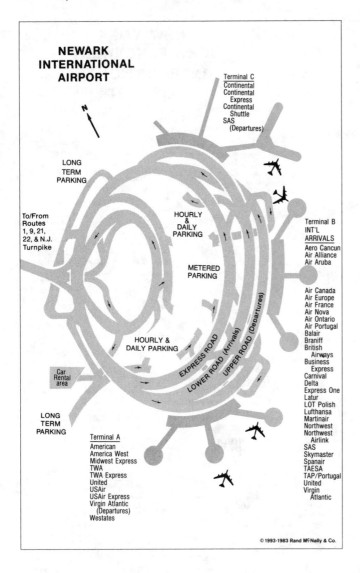

NEWARK INTERNATIONAL AIRPORT

Terminal C
Continental
Continental
Express
Continental
Shuttle
SAS
(Departures)

LONG TERM PARKING

HOURLY & DAILY PARKING

To/From Routes 1, 9, 21, 22, & N.J. Turnpike

METERED PARKING

Terminal B
INT'L ARRIVALS
Aero Cancun
Air Alliance
Air Aruba

Air Canada
Air Europe
Air France
Air Nova
Air Ontario
Air Portugal
Balair
Braniff
British Airways
Business Express
Carnival
Delta
Express One
Latur
LOT Polish
Lufthansa
Martinair
Northwest
Northwest Airlink
SAS
Skymaster
Spanair
TAESA
TAP/Portugal
United
Virgin Atlantic

HOURLY & DAILY PARKING

EXPRESS ROAD
LOWER ROAD (Arrivals)
UPPER ROAD (Departures)

Car Rental area

LONG TERM PARKING

Terminal A
American
America West
Midwest Express
TWA
TWA Express
United
USAir
USAir Express
Virgin Atlantic
(Departures)
Westates

© 1993-1983 Rand McNally & Co.

SELECTED RESTAURANTS:

The Four Seasons, 99 E. 52nd St., 754-9494

Fraunces Tavern Restaurant, 54 Pearl St., 269-0144

Gallagher's Steak House, 228 W. 52nd, 245-5336

La Cote Basque, 5 E. 55th St., 688-6525

Lutèce, 249 E. 50th St., 752-2225

Mitsukoshi, 461 Park Ave., 935-6444

The Russian Tea Room, 150 W. 57th, 265-0947

Sardi's, 234 W. 44th, 221-8440

Stage Delicatessen, 834 7th Ave., 245-7850

Toots Shore, 233 W. 33rd, 630-0333

"21" Club, 21 W. 52nd St., 582-7200

Windows on the World, 1 World Trade Center—West St. entrance, 938-1111

SELECTED ATTRACTIONS:

American Museum of Natural History, 79th & Central Park W., 769-5100

Empire State Building, Fifth Ave. & 34th St., 736-3100

Guggenheim Museum, Fifth Ave. at 89th St., 423-3600

Intrepid Sea-Air-Space Museum, W. 46th St. & 12th Ave., 245-0072

Metropolitan Museum of Art, Fifth Ave. & 82nd St., 535-7710

Museum of Modern Art, 11 W. 53rd St., 708-9480

NBC Tours, 30 Rockefeller Plaza, 664-7174

South Street Seaport, Water & Fulton sts., 732-7678

Statue of Liberty, Liberty Island, 363-3200

World Trade Center Observation Deck, 2 World Trade Center, 435-7000

INFORMATION SOURCE:

New York Convention and Visitors Bureau, Inc.
   Two Columbus Circle
   New York City, New York 10019
   (212) 397-8200

Norfolk-Virginia Beach, Virginia

Population: (*1,396,107 Norfolk-Virginia Beach-Newport News Metro Area) Norfolk 261,229;
   Virginia Beach 393,069 (1990C)
Altitude: Sea level to 12 feet
Average Temp.: Jan., 41°F.; July, 77°F.
Telephone Area Code: 804
Time: 622-9311   **Weather:** 666-1212
Time Zone: Eastern

AIRPORT TRANSPORTATION:

Ten miles to downtown Norfolk; twenty-two miles to downtown Virginia Beach.

Taxicab, limousine and bus service to Norfolk; taxicab and airport limousine service to Virginia Beach.

SELECTED HOTELS: NORFOLK

Airport Hilton, 1500 N. Military Hwy., 466-8000

Holiday Inn Executive Center, 5655 Greenwich Rd., 499-4400

Hotel Norfolk, 700 Monticello Ave., 627-5555

Omni International Hotel, 777 Waterside Dr., 622-6664

Ramada Inn, 6360 Newtown Rd., 461-1081

Ramada Madison Hotel, Granby & Freemason sts., 622-6682

Sheraton Inn—Military Circle, 870 N. Military Hwy., 461-9192

Waterside Marriott Hotel, Main & Atlantic sts. 627-4200

SELECTED RESTAURANTS: NORFOLK

Cafe Charlieu, 112 College Place off Boush St., 623-7202

La Galleria, 120 College Pl., 623-3939

Lockhart's Seafood Restaurant, 8440 Tidewater Dr., 588-0405

Riverwalk Cafe, in the Omni International Hotel, 622-6664

The Ship's Cabin Seafood Restaurant, 4110 E. Ocean View Ave., 480-2526

SELECTED ATTRACTIONS: NORFOLK

American Rover Tall Sailing Ship Tours, 627-SAIL

Carrie B Harbor Tours, 393-4735

Chrysler Museum of Art, Olney Rd. & Mowbray Arch, 622-1211

Douglas MacArthur Memorial Museum, City Hall Ave. & Bank St., 441-2965

Hermitage Foundation Museum, 7637 N. Shore Rd., 423-2052

Nauticus—The National Maritime Center, west end of Main St. at the Waterfront, 623-9084

Norfolk Botanical Gardens, Azalea Garden Rd., 441-5830

Norfolk Naval Base (tour), 9809 Hampton Blvd., 444-7955

Virginia Zoological Park, 3500 Granby St., 441-2706

The Waterside (festival marketplace), 333 Waterside Dr., 627-3300

INFORMATION SOURCES: NORFOLK

Norfolk Convention and Visitors Bureau
   236 E. Plume St.
   Norfolk, Virginia 23510
   (804) 441-5266

Hampton Roads Chamber of Commerce
420 Bank St.
Norfolk, Virginia 23510
(804) 622-2312

SELECTED HOTELS: VIRGINIA BEACH
Colonial Inn Motel, 29th & Oceanfront, 428-5370
Comfort Inn, 2800 Pacific Ave., 428-2203
Courtyard by Marriott, 5700 Greenwich Rd., 490-2002
Days Inn Airport, 5708 Northampton Blvd., 460-2205
Founders Inn & Conference Center, I-64 & Indian River Rd., 424-5511
Holiday Inn Airport, US 13 & I-64, 464-9351
Holiday Inn Executive Center, 5655 Greenwich Rd. & Newtown Rd. Exit, 499-4400
Holiday Inn on the Ocean, 39th & Oceanfront, 428-1711
Quality Inn Pavilion, Parks Ave. at 21st, 422-3617
Radisson Hotel Virginia Beach, 1900 Pavilion Dr., 422-8900

SELECTED RESTAURANTS: VIRGINIA BEACH
Alexander's on the Bay, Fentress St. at Chesapeake Bay, 464-4999
Beach Pub, 1001 Laskin Rd., 422-8817
Belle Monte Cafe, 134 Hilltop E., 425-6290
Bennigan's, 757 Lynnhaven Pkwy., 463-7100
Casual Clam, 3101 Virginia Beach Blvd., 463-5106
Joe's Sea Grill, 981 Laskin Rd., 422-5637
Lista's Mexican Restaurant & Cantina, 3900 Bonney Rd., 463-8226
Olive Garden, 6831 Lynnhaven Pkwy., 486-8234
Orion's, in the Cavalier Hotel, 42nd & Oceanfront, 425-8555
Szechuan Garden, 2720 N. Mall Dr., 463-1680
Three Ships Inn, 3800 Shore Dr., 460-0055

SELECTED ATTRACTIONS: VIRGINIA BEACH
Adam Thoroughgood House (17th-century), 1636 Parish Rd., 460-0007
Association for Research & Enlightenment (headquarters for the work of psychic Edgar Cayce), 67th St. & Atlantic Ave., 428-3588
Atlantic Fun Center, 25th St. & Atlantic Ave., 422-1742
Back Bay National Wildlife Refuge, 4005 Sandpiper Rd., 721-2412
Christian Broadcasting Network, I-64 at Indian River Rd., 523-7123
First Landing Cross, Fort Story
Lynnhaven House (circa 1725), 4405 Wishart Rd., 460-1688
Ocean Breeze Amusement Park, 849 Central Booth Blvd., 422-4444; 340-1616 off-season
Seashore State Park, 2500 Shore Dr., 481-4836
Virginia Marine Science Museum, 717 General Booth Blvd., 437-4949 or 425-FISH (recording)

INFORMATION SOURCES: VIRGINIA BEACH
Virginia Beach Convention & Visitors Bureau
2101 Parks Ave., Suite 500
Virginia Beach, Virginia 23451
(804) 437-4700
Virginia Beach Visitor Information Center
2100 Parks Ave.
P.O. Box 200
Virginia Beach, Virginia 23458
(804) 437-4888; (800) 446-8038
Hampton Roads Chamber of Commerce
4512 Virginia Beach Blvd.
Virginia Beach, Virginia 23462
(804) 490-1223

Oklahoma City, Oklahoma

Population: (*958,839) 444,719 (1990C)
Altitude: 1,243 feet
Average Temp.: Jan., 37°F.; July, 82°F.
Telephone Area Code: 405
Time: 599-1234 **Weather:** 524-3377
Time Zone: Central

AIRPORT TRANSPORTATION:
Ten miles to downtown Oklahoma City.
Taxicab, limousine bus service.

SELECTED HOTELS:
Days Inn, 2616 S. I-35, 677-0521
Days Inn, 2801 NW. 39th St., 946-0741
Embassy Suites, 1815 S. Meridian, 682-6000
Hilton Inn Northwest, 2945 NW. Expwy., 848-4811

Hilton Inn West, 401 S. Meridian Ave., 947-7681
Holiday Inn East, 5701 Tinker Diagonal, Midwest City, 737-4481
Lincoln Plaza Hotel & Conference Center, 4445 N. Lincoln Blvd., 528-2741
Oklahoma City Marriott, 3233 NW. Expwy., 842-6633
Sheraton-Century Center Hotel & Towers, One N. Broadway Ave., 235-2780
Waterford Hotel, 6300 Waterford Blvd., 848-4782

SELECTED RESTAURANTS:
Applewoods, 4301 SW. 3rd, 947-8484
Eagle's Nest, 5900 Mosteller Dr., Top floor, 840-5655
Eddy's Steak House, 4227 N. Meridian Ave., 787-2944
Harry Bear's American Grill, 4540 NW. 23rd St., 946-1421
Harry Bear's II, 5705 Mosteller Dr., 840-9912
Oklahoma County Line, 1226 NE. 63rd, 478-4955
Shorty Small's, 4500 W. Reno, 947-0779
Sleepy Hollow, 1101 NE. 50th, 424-1614
Sunshine Express, in the Sheraton-Century Center Hotel & Towers, 235-2780
Texanna Red's, 4600 W. Reno Ave., 947-8665

SELECTED ATTRACTIONS:
Enterprise Square, 2501 E. Memorial Rd., 425-5030
Frontier City (theme park), 11501 NE Expressway, 478-2414

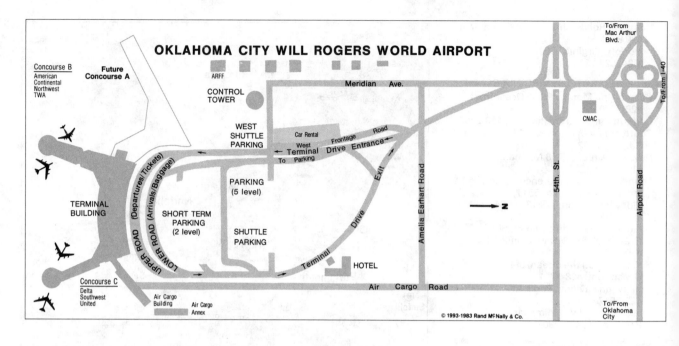

OKLAHOMA CITY WILL ROGERS WORLD AIRPORT

© 1993-1983 Rand McNally & Co.

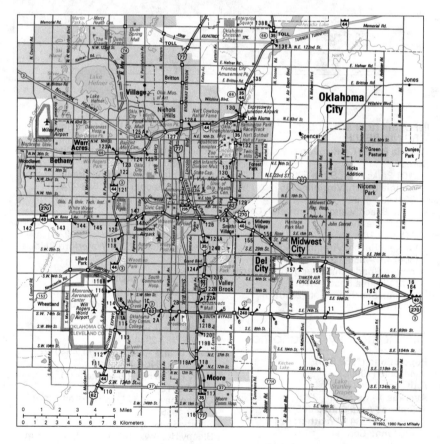

© 1992, 1980 Rand McNally

INFORMATION SOURCES:
Oklahoma City Convention & Visitors Bureau
4 Santa Fe Plaza
Oklahoma City, Oklahoma 73102
(405) 278-8912
Oklahoma City Chamber of Commerce
One Santa Fe Plaza
Oklahoma City, Oklahoma 73102
(405) 278-8900

Omaha, Nebraska

Population: (*618,262)
335,795 (1990C)
Altitude: 1,040 feet
Average Temp.: Jan., 23°F.; July, 77°F.
Telephone Area Code: 402
Time: 342-8463   **Weather:** 392-1111
Time Zone: Central

AIRPORT TRANSPORTATION:
Three miles to downtown Omaha.
Taxicab, bus and hotel limousine service.

SELECTED HOTELS:
Best Western Airport Inn, at Eppley Airfield, 348-0222
Best Western New Tower Inn, 7764 Dodge St., 393-5500
Best Western Omaha Inn, 4706 S. 108th St., 339-7400
Best Western Regency West, 107th and Pacific sts., 397-8000
Embassy Suites, 7270 Cedar St., 397-5141
Holiday Inn Central, 3321 S. 72nd St., 393-3950
Howard Johnson's Omaha, 3650 S. 72nd., 397-3700
Omaha Marriott, 10220 Regency Circle, 399-9000
Ramada—Central, 7007 Grover St., 397-7030
Ramada Inn Airport, Abbott Dr. & Locust St., 342-5100

Horse Shows (year-round), at the State Fairgrounds, 948-6700
Kirkpatrick Center Complex, 2100 NE 52nd, 427-5461
Myriad Gardens/Crystal Bridge, Reno & Robinson, 297-3995
National Cowboy Hall of Fame & Western

Heritage Center, 1700 NE 63rd St., 478-2250
Oklahoma City Zoo, NE 50th & Martin Luther King Blvd., 424-3344
Remington Park (racetrack), One Remington Pl., 424-9000

Red Lion Inn Omaha, 1616 Dodge St.,
   346-7600
Sheraton Inn Southwest, 4888 S. 118th
   St., 895-1000

SELECTED RESTAURANTS:
Anthony's, 7220 F St., 331-7575
Cascio, 1622 10th St., 345-8313
Chardonnay, in the Omaha Marriott,
   399-9000
French Cafe, 1017 Howard St., 341-3547
Gallagher's, 10730 Pacific, 393-1421
Gorats Steak House, 4917 Center St.,
   551-3733
Johnny's, 4702 S. 27th St., 731-4774
Maxine's, in the Red Lion Inn Omaha
   Hotel, 346-7600
Neon Goose, 1012 S. 10th St., 341-2063
Ross' Steak House, 909 S. 72nd St.,
   393-2030
V. Mertz, 1022 Howard, 345-8980

SELECTED ATTRACTIONS:
Ak-Sar-Ben Thoroughbred Racetrack,
   63rd & Shirley, 556-2305
Belle at Bellevue Riverboat Cruises,
   Haworth Park, Bellevue, 292-BOAT
Boys Town, W. Dodge Rd. between
   132nd & 144th, 498-1140
Central Park Mall, 14th & Farnam
Heartland of America Park & Fountain,
   downtown by Central Park Mall
Henry Doorly Zoo, 3701 S. 10th, 733-8400
Joslyn Art Museum, 24th & Dodge,
   342-3300
Old Market, 10th to 13th & Howard sts.
Peony Park, 81st & Cass
Strategic Air Command (SAC) Museum,
   2510 Clay St., Bellevue, 292-2001

INFORMATION SOURCES:
Greater Omaha Convention & Visitors
Bureau
   1819 Farnam, Suite 1200
   Omaha, Nebraska 68183
   (402) 444-4660
Greater Omaha Chamber of Commerce
   1301 Harney St.
   Omaha, Nebraska 68102
   (402) 346-5000

Orlando, Florida

Population: (*1,072,748)
   164,693 (1990C)
Altitude: 106 feet
Average Temp.: Jan., 62°F.; July, 82°F.
Telephone Area Code: 407
Time: 422-1611 **Weather:** 851-7510
Time Zone: Eastern

AIRPORT TRANSPORTATION:
Fifteen miles to downtown Orlando.
Taxicab, limousine and bus service.

SELECTED HOTELS:
Delta Court of Flags Hotel, 5715 Major
   Blvd., 351-3340
Gold Key Inn, 7100 S. Orange Blossom
   Trail, 855-0050
Holiday Inn—Central Park, 7900 S.
   Orange Blossom Trail, 859-7900
Holiday Inn—Centroplex, 929 W. Colonial
   Dr., 843-1360
Howard Johnson's Downtown, 304 W.
   Colonial Dr. at jct. FL 50 & I-4, 843-8700

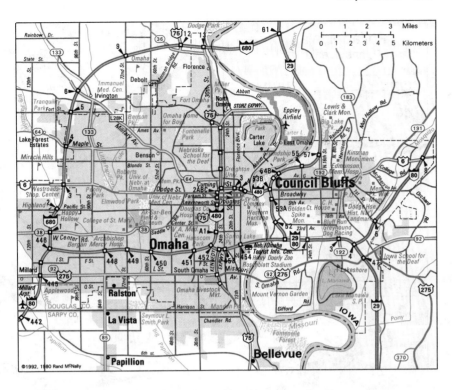

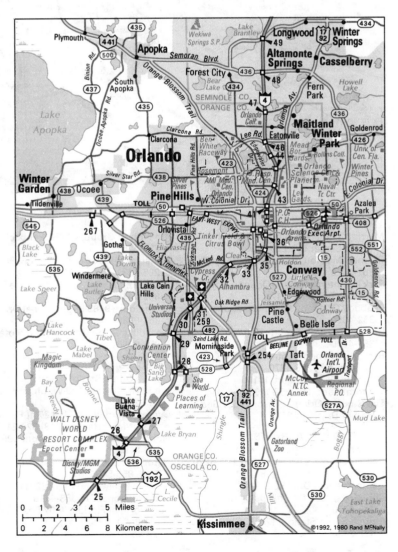

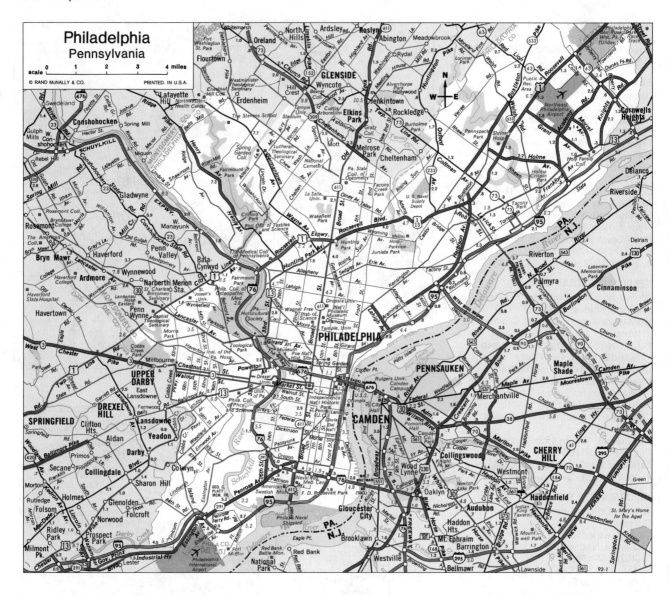

Philadelphia
Pennsylvania
scale 0 1 2 3 4 miles
© RAND McNALLY & CO.   PRINTED IN U.S.A.

Marriott's Orlando World Center, 8701
World Center Dr., 239-4200
Orlando Marriott, 8001 International Dr.,
351-2420
Park Inn International—Orlando North,
736 Lee Rd., 647-1112
Peabody Orlando, 9801 International Dr.,
352-4000
Ramada Inn, 7400 International Dr.,
351-4600
Ramada Orlando Central, 3200 W.
Colonial Dr., 295-5270
Sheraton World Resort, 10100
International Dr., 352-1100
The Stouffer Orlando Resort, 6677 Sea
Harbor Dr., 351-5555

SELECTED RESTAURANTS:
Cafe on the Park, in the Harley Hotel of
Orlando, 151 E. Washington, 841-3220
Charlie's Lobster House, 8445
International Dr., 352-6929
Christini's, 7600 Dr. Phillips Blvd.,
345-8770
Church Street Station, 129 W. Church St.,
422-2434

4th Fighter Group, 494 Rickenbacker Dr.,
898-4251
House of Beef, 801 John Young Pkwy.,
295-1931
Maison et Jardin, 430 S. Wymore Rd.,
862-4410
Ming Court, 9188 International Dr.,
351-9988
Piccadilly, in the Gold Key Inn Motel,
855-0050

SELECTED ATTRACTIONS:
Church Street Station (dining, shopping,
entertainment), 129 W. Church St.,
422-2434
Cypress Gardens, off US 27 at FL 540W,
near Winter Haven, (813) 324-2111
Florida Citrus Tower, Clermont, (904)
394-8585
Gatorland Zoo, 14501 S. Orange Blossom
Trail, 855-5496
Ripley's Believe It or Not, 8201
International Dr., 363-4418
Sea World of Florida, 7007 Sea World Dr.,
351-3600

Spaceport USA, Kennedy Space Center,
452-2121
Universal Studios Florida, 1000 Universal
Studio Plaza, 363-8000
Walt Disney World, Lake Buena Vista,
824-4321

INFORMATION SOURCES:
Orlando/Orange County Convention and
Visitors Bureau
7208 Sand Lake Rd., Suite 300
Orlando, Florida 32819
(407) 363-5800
Greater Orlando Chamber of Commerce
75 E. Ivanhoe Blvd.
Orlando, Florida 32802
(407) 425-1234

Philadelphia, Pennsylvania

Population: (*4,856,881)
1,585,577 (1990C)
Altitude: 45 feet
Average Temp.: Jan., 35°F.; July, 78°F.
Telephone Area Code: 215

Time: 846-1212   **Weather:** 936-1212
Time Zone: Eastern

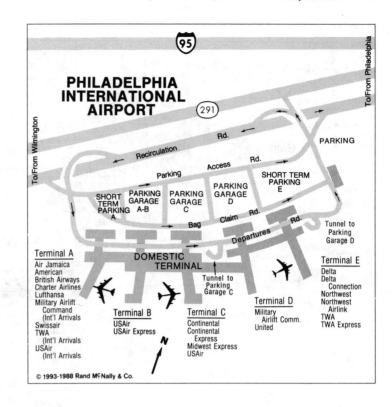

PHILADELPHIA INTERNATIONAL AIRPORT

© 1993-1988 Rand McNally & Co.

AIRPORT TRANSPORTATION:

Eight miles to downtown Philadelphia.
Taxicab, limousine bus, and rail line
service.

SELECTED HOTELS:

Adam's Mark, City Ave. & Monument Rd.,
581-5000
The Barclay Hotel, 237 S. 18th St.,
545-0300
The Four Seasons, 18th St. & Benjamin
Franklin Pkwy., 963-1500
Hotel Atop the Bellevue, Broad & Walnut
sts., 893-1776
The Latham Hotel, 17th & Walnut sts.,
563-7474
Penn Tower Hotel, 34th & Civic Center
Blvd., 387-8333
Philadelphia Airport Marriott Hotel, 4509
Island Ave., 365-4150
Ramada Inn, 2400 Old Lincoln Hwy.,
Trevose, 638-8300
Sheraton Society Hill, 1 Dock St.,
238-6000
The Warwick Hotel, 17th at Locust St.,
735-6000
Wyndham Franklin Plaza Hotel, 17th &
Race, 448-2000

SELECTED RESTAURANTS:

Deja Vu, 1609 Pine St., 546-1190
Deux Cheminees, 1221 Locust St.,
790-0200
Di Lullo Centro, 1407 Locust, 546-2000
La Famiglia, 8 S. Front St., 922-2803
Lautrec, 408 S. Second St., 923-6660
Le Bec Fin, 1523 Walnut St., 567-1000
The Monte Carlo Living Room, 2nd &
South sts., 925-2220
Old Original Bookbinders, 125 Walnut St.,
925-7027

SELECTED ATTRACTIONS:

Academy of Natural Sciences, 19th &
Benjamin Franklin Pkwy., 299-1066
Afro-American Historical & Cultural
Museum, NW corner of 7th & Arch sts.,
574-0380
Betsy Ross House, 239 Arch St., 627-5343
Franklin Institute-Futures Center &
Omniverse Theater, 20th & Benjamin
Franklin Pkwy., 448-1208
Independence National Historical Park,
3rd & Chestnut (Vistors Center),
597-8974
The New Jersey State Aquarium at
Camden, Riverside Dr., Camden, New
Jersey, (609) 365-3300
Philadelphia Museum of Art, 26th &
Benjamin Franklin Pkwy., 763-8100
Philadelphia Zoo, 34th & Girard Ave.,
243-1100
Please Touch Museum for Children, 210
N. 21st St., 963-0666
U.S. Mint, 5th & Arch sts., 597-7353

INFORMATION SOURCES:

Philadelphia Convention & Visitors
Bureau
    1515 Market St., Suite 2020
    Philadelphia, Pennsylvania 19102
    (215) 636-3300

Greater Philadelphia Chamber of
Commerce
    1234 Market St., 18th Floor
    Philadelphia, Pennsylvania 19107
    (215) 545-1234

Phoenix, Arizona

Population: (*2,122,101)
    983,403 (1990C)
Altitude: 1,090 feet
Average Temp.: Jan., 51°F.; July, 85.9°F.
Telephone Area Code: 602
Time and Weather: 1-976-7600
Time Zone: Mountain Standard all year

AIRPORT TRANSPORTATION:

Four miles to downtown Phoenix.
Taxicab, bus, and limousine bus service.

SELECTED HOTELS:

Arizona Biltmore, 24th St. & Missouri
Ave., 955-6600
Embassy Suites Hotel, 2333 E. Thomas
Rd., 957-1910
Hotel Park Central, 3600 N. 2nd Ave.,
248-0222
Hyatt Regency Phoenix, 122 N. 2nd St.,
252-1234
Omni Adam, Central & Adams sts.,
257-1525
Orange Tree Golf & Conference Resort,
10601 N. 56th St., Scottsdale, 948-6100
The Phoenix Airport Hilton, 2435 S. 47th
St., 894-1600
The Pointe Resort, 7677 N. 16th St.,
997-2626
Ritz-Carlton Phoenix, 24th St. &
Camelback Rd., 468-0700

SELECTED RESTAURANTS:

Compass, in the Hyatt Regency Phoenix,
252-1234
Etienne's Different Pointe of View, 11111
N. 7th St., 863-0912
The Hungry Hunter, 3102 E. Camelback
Rd., 957-7180
Orangerie, in the Arizona Biltmore Hotel,
954-2507
Trumps, in the Hotel Westcourt, 10220 N.
Metro Pkwy. E., 997-5900

SELECTED ATTRACTIONS:

Arabian Productions & Tours, 18001 N.
Tatum Blvd., Scottsdale, 867-8275
Champlin Fighter Museum, 4636 Fighter
Aces Dr., Mesa, 830-4540
Desert Botanical Gardens, 1201 N. Galvin
Pkwy., 941-1225
Dolly's Steamboat (Canyon Lake tours),
827-9144
Gila River Arts & Crafts Center, Sacaton,
963-3981
Hall of Flame Museum, Project Dr.,
275-3473
The Heard Museum, 22 E. Monte Vista,
252-8840
Pueblo Grande Museum, 4619 E.
Washington, 495-0901
Rawhide (recreated 1880s Western town),
Scottsdale, 563-5111
Taliesin West, Scottsdale, 860-8810

INFORMATION SOURCE:

Phoenix & Valley of the Sun Convention
& Visitors Bureau
    One Arizona Center
    400 E. Van Buren St., Suite 600
    Phoenix, Arizona 85004-2290
    (602) 254-6500
    Visitor Info. Hotline (602) 252-5588

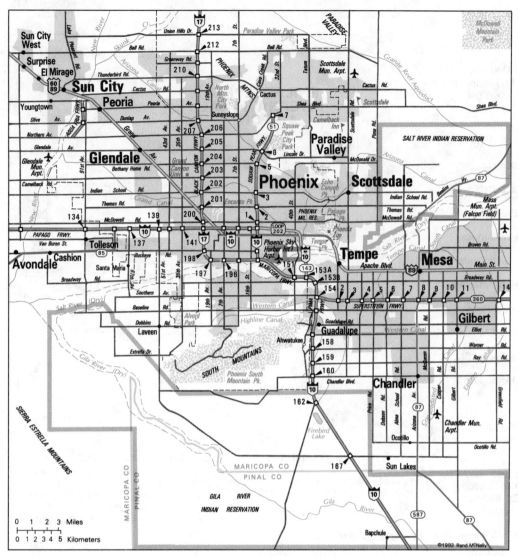

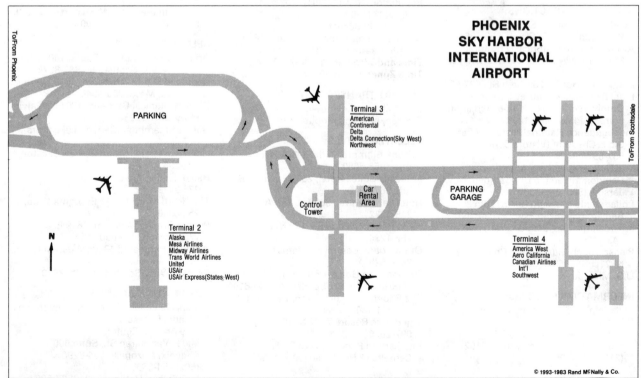

PHOENIX SKY HARBOR INTERNATIONAL AIRPORT

To/From Phoenix

To/From Scottsdale

PARKING

Terminal 3
American
Continental
Delta
Delta Connection(Sky West)
Northwest

Car Rental Area

PARKING GARAGE

Control Tower

N

Terminal 2
Alaska
Mesa Airlines
Midway Airlines
Trans World Airlines
United
USAir
USAir Express(States West)

Terminal 4
America West
Aero California
Canadian Airlines Int'l
Southwest

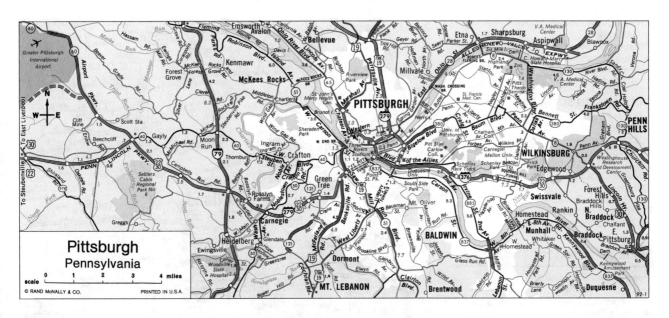

Pittsburgh, Pennsylvania

Population: (*2,056,705)
369,879 (1990C)
Altitude: 760 feet
Average Temp.: Jan., 33°F.; July, 75°F.
Telephone Area Code: 412
Time: 391-9500   **Weather:** 936-1212
Time Zone: Eastern

AIRPORT TRANSPORTATION:

Seventeen miles to downtown
Pittsburgh.
Taxicab and limousine bus service.

SELECTED HOTELS:

Holiday Inn Airport—Pittsburgh, 1406
Beers School Rd., Coraopolis, 262-3600
Hyatt Pittsburgh at Chatham Center, 112
Washington Pl., 471-1234
Marriott—Greentree, 101 Marriott Dr.,
922-8400
Pittsburgh Airport Hilton Inn, Parkway
West at the Montour Run Exit, 262-3800
The Pittsburgh Hilton & Towers, 600
Commonwealth Pl., 391-4600
Royce Hotel, 1160 Thorn Run Rd.
Extension, Coraopolis, 262-2400
Sheraton Hotel at Station Square, Carson
& Smithfield sts., 261-2000
Vista International Hotel, 1000 Penn Ave.,
281-3700
Westin William Penn, 530 William Penn
Pl., 281-7100

SELECTED RESTAURANTS:

Christopher's, 1411 Grandview Ave., Mt.
Washington, 381-4500
Colony, Greentree & Cochran rds.,
561-2060
Common Plea, 310 Ross St., 281-5140
D'Imperio's, 3412 Wm. Penn Hwy.,
823-4800
Grand Concourse, 1 Station Sq., 261-1717
Le Mont Restaurant, 1114 Grandview
Ave., 431-3100
Tambellini, S. of town on PA 51, 481-1118
The Terrace Room, in the Westin William
Penn, 281-7100

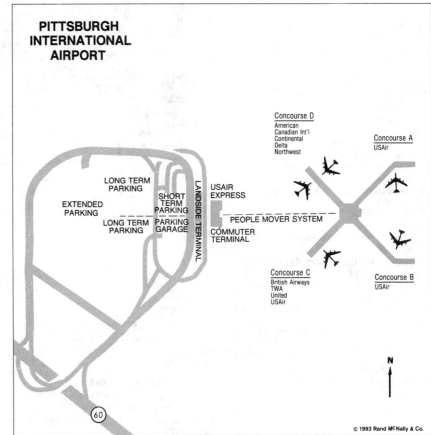

Top of the Triangle, 600 Grant St.,
471-4100

SELECTED ATTRACTIONS:

Benedum Center for the Performing Arts,
Penn Ave. & 7th St., 456-2600
Carnegie Museum of Art & Natural
History, 4400 Forbes Ave., 622-3172

Carnegie Science Center, 1 Allegheny
Ave., 237-3300
Frank Lloyd Wright's Fallingwater,
between Mill Run & Ohiopyle on PA
381, 329-8501
Heinz Hall (Pittsburgh Symphony), 600
Penn Ave., 392-4800
Monongahela Incline (cable car), 205 W.
Carson St., South Side, 231-5707

Phipps Conservatory, Schenley Park, Oakland, 622-6914
Pittsburgh Zoo, Highland Park, 665-3639
Station Square, W. Carson St. & Smithfield St. bridge, South Side, 261-9911

INFORMATION SOURCES:
Greater Pittsburgh Convention & Visitors Bureau
4 Gateway Center
Pittsburgh, Pennsylvania 15222
(412) 281-7711; (800) 366-0093
Greater Pittsburgh Chamber of Commerce
3 Gateway Center
Pittsburgh, Pennsylvania 15222
(412) 392-4500

Portland, Oregon

Population: (*1,239,842) 437,319 (1990C)
Altitude: Sea level to 1,073 feet
Average Temp.: Jan., 40°F.; July, 69°F.
Telephone Area Code: 503
Time: 1-976-1111 **Weather:** 281-1911
Time Zone: Pacific

AIRPORT TRANSPORTATION:
Nine miles to downtown Portland.
Taxicab, limousine bus service and public mass transit system.

SELECTED HOTELS:
The Benson, 309 SW. Broadway, 228-2000
Best Western Flamingo, 9727 NE. Sandy Blvd., 255-1400
The Heathman Hotel, 1009 SW. Broadway, 241-4100
Holiday Inn—Portland Airport, 8439 NE. Columbia Blvd., 256-5000
Portland Hilton, 921 SW. 6th Ave., 226-1611
The Portland Inn, 1414 SW. 6th Ave., 221-1611
Portland Marriott, 1401 SW. Front Ave., 226-7600
Ramada Inn at the Airport, 6221 NE. 82nd Ave., 255-6511
Red Lion Coliseum, 1225 N. Thunderbird Way, 235-8311
Red Lion Columbia River, 1401 N. Hayden Island Dr., 283-2111
Red Lion Inn Downtown, 310 SW. Lincoln Ave., 221-0450
Red Lion Motor Inn Jantzen Beach, 909 N. Hayden Island Dr., 283-4466
Sheraton Inn—Portland Airport, 8235 NE. Airport Way, 281-2500

SELECTED RESTAURANTS:
Alexander's, atop Portland Hilton, 226-1611
Brickstone's Restaurant, in the Red Lion Columbia River, 283-2111
Couch Street Fish House, 105 NW. 3rd St., 223-6173
Huber's, 411 SW. 3rd., 228-5686
Jake's Famous Crawfish Restaurant, 401 SW. 12th Ave., 226-1419
The London Grill, in The Benson Hotel, 228-2000
Ocean Palace, 140 NW. 4th Ave., 223-7475

Polo Club, 718 NE. 12th Ave., 232-1801
Ringside, 2165 W. Burnside St., 223-1513
River Queen, 1300 NW. Front Ave., 228-8633

SELECTED ATTRACTIONS:
International Rose Test Gardens, in Washington Park, 796-5193
Japanese Gardens, in Washington Park, 223-4070
Oregon Historical Society's Library & Museum, 1230 SW. Park Ave., 222-1741
Oregon Museum of Science & Industry, 4015 SW. Canyon Rd., 222-2828
Pittock Mansion, 3229 NW. Pittock Dr., 823-3624
Portland Art Museum, 1219 SW. Park Ave., 226-2811
Sanctuary of Our Sorrowful Mother (the "Grotto"), NE. 85th & Sandy Blvd., 254-7371

Washington Park Zoo, 4001 SW. Canyon Rd., 226-1561
World Forestry Center, 4033 SW. Canyon Rd. in Washington Park, 228-1367

INFORMATION SOURCES:
Portland/Oregon Visitors Association Marketing, Tourism & Conventions
26 SW. Salmon St.
Portland, Oregon 97204
(503) 275-9750
Portland Chamber of Commerce
221 NW. 2nd Ave.
Portland, Oregon 97209
(503) 228-9411

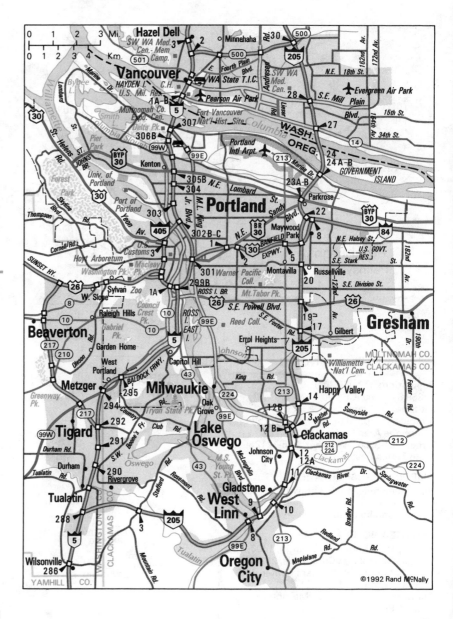

©1992 Rand McNally

Providence, Rhode Island

Population: (*654,854)
 160,728 (1990C)
Altitude: 24 feet
Average Temp.: Jan., 29°F.; July, 72°F.
Telephone Area Code: 401
Time: 776-2700 **Weather:** 737-5100
Time Zone: Eastern

AIRPORT TRANSPORTATION:
Nine miles to downtown Providence.
Taxicab and airport limousine service.

SELECTED HOTELS:
Holiday Inn Downtown, 21 Atwells Ave.,
 831-3900
Inn at the Crossing—Holiday Inn, 801
 Greenwich Ave., Warwick, 732-6000
Omni Biltmore, Kennedy Plaza, 11
 Dorrance St., 421-0700
Providence Marriott, Charles & Orms sts.,
 272-2400
Ramada Inn, 940 Fall River Ave.,
 Seekonk, MA, (508) 336-7300

SELECTED RESTAURANTS:
Camille's, 71 Bradford St., 751-4812
Capriccio, Dyer and Pine sts., 421-1320
Old Grist Mill Tavern, 390 Fall River Ave.,
 Seekonk, MA, (508) 336-8460
Raphael's, 207 Pine St., 421-4646
Stacey's, in the Providence Marriott,
 272-2400

SELECTED ATTRACTIONS:
The Arcade, 65 Weybosset St., 272-2340
Benefit Street ("Mile of History"),
 831-7440
First Baptist Meeting House, 75 N. Main
 St., 421-1177
Governor Stephen Hopkins House,
 Benefit & Hopkins sts., 884-8337
John Brown House, 52 Power St.,
 331-8575
Lippitt House, 199 Hope St., 453-0688
Museum of Art, Rhode Island School of
 Design, 224 Benefit St., 454-6500 or
 -6507
Roger Williams National Memorial, 282
 N. Main St., 528-5385
State House (tours), 82 Smith St.,
 277-2357

INFORMATION SOURCES:
Greater Providence Convention &
Visitors Bureau
 30 Exchange Terrace
 Providence, Rhode Island 02903
 (401) 274-1636
Greater Providence Chamber of
Commerce
 30 Exchange Terrace
 Providence, Rhode Island 02903
 (401) 521-5000

Raleigh, North Carolina

Population: (*735,480)
 207,951 (1990C)
Altitude: 363 feet
Average Temp.: Jan., 42°F.; July, 78°F.
Telephone Area Code: 919
Time: 976-2511 **Weather:** 840-0450
Time Zone: Eastern

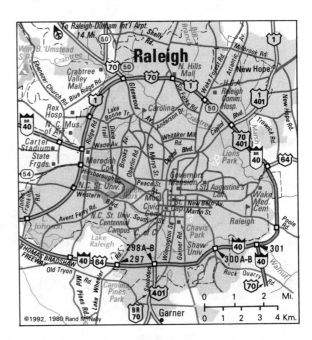

©1992, 1980 Rand McNally

AIRPORT TRANSPORTATION:
Fifteen miles to downtown Raleigh.
Taxicab and airport limousine service.

SELECTED HOTELS:
Brownstone Hotel, 1707 Hillsborough St.,
 828-0811
Days Inn—North, 2805 Highwood Blvd.,
 872-3500
Holiday Inn—North, 2815 Capital Blvd.,
 872-7666
Holiday Inn—Raleigh State Capital, 320
 Hillsborough St., 832-0501
North Raleigh Hilton Convention Center
 and Towers, 3415 Wake Forest Rd.,
 872-2323
Plantation Inn, 6401 Capital Blvd.,
 876-1411
Radisson Plaza Hotel Raleigh, 420
 Fayetteville St. Mall, 834-9900
Raleigh Marriott Hotel, 4500 Marriott Dr.,
 781-7000
Ramada Inn—South, on US 1 at jct. NC
 55, 362-8621
Sheraton—Crabtree Inn, 4501
 Creedmoor Rd., 787-7111
Velvet Cloak Inn, 1505 Hillsborough St.,
 828-0333

SELECTED RESTAURANTS:
Allies, in the Raleigh Marriott Hotel,
 781-7000
Angus Barn Ltd., 9401 Glenwood Ave.,
 787-3505
Jacqueline's, in the Plantation Inn,
 876-1411
Jeremiah's, in the Mission Valley Inn,
 2110 Avent Ferry Rd., 828-3173
Top of the Tower, in the Holiday Inn—
 Raleigh State Capital, 832-0501

SELECTED ATTRACTIONS:
Artspace, 201 E. Davie St., 821-2787
City Cemetery (1798), S. East & Hargett
 sts.

Executive Mansion, 200 N. Blount St.,
 733-3456
Hardee's Pavilion Walnut Creek
 Amphitheatre, 3801 Rock Quarry Rd.,
 831-6400
Mordecai Historic Park, 1 Mimosa St.,
 834-4844
North Carolina Museum of Art, 2110 Blue
 Ridge Blvd., 833-1935
North Carolina Museum of History, 109 E.
 Jones St., 733-3894
Oakwood Cemetery (1866), Oakwood
 Ave. at Watauga St.
State Capitol, bounded by Wilmington,
 Edenton, Salisbury, and Morgan sts.,
 733-4994
Wakefield (oldest dwelling in Raleigh),
 Hargett & St. Mary's sts., 733-3456

INFORMATION SOURCES:
Greater Raleigh Convention & Visitors
Bureau
 225 Hillsborough St., Suite 400
 P.O. Box 1879
 Raleigh, North Carolina 27602
 (919) 834-5900; (800) 849-8499
Greater Raleigh Chamber of Commerce
 800 S. Salisbury St.
 P.O. Box 2978
 Raleigh, North Carolina 27602
 (919) 664-7000

Richmond, Virginia

Population: (*865,640)
 203,056 (1990C)
Altitude: 150 feet
Average Temp.: Jan., 38°F.; July, 78°F.
Telephone Area Code: 804
Time: 844-3711 **Weather:** 268-1212
Time Zone: Eastern

AIRPORT TRANSPORTATION:
Ten miles to downtown Richmond.
Taxicab and limousine service.

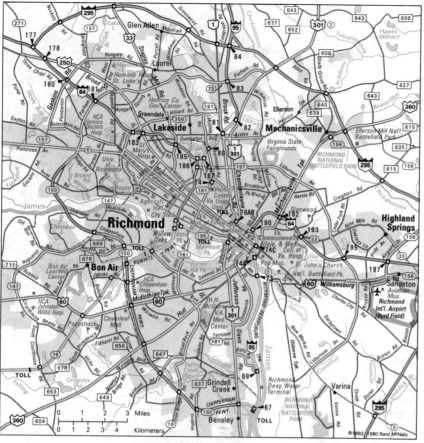

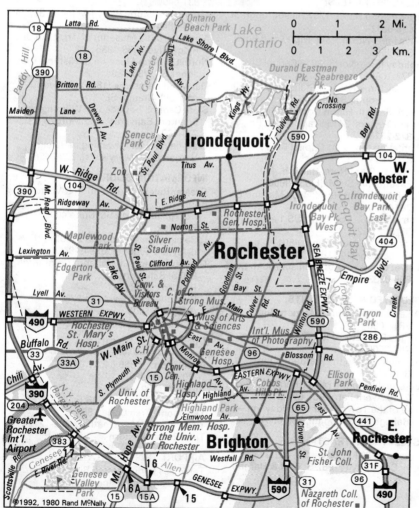

SELECTED HOTELS:

Commonwealth Park Suites Hotel, 9th & Bank sts., 343-7300

Courtyard by Marriott, 6400 W. Broad St., 282-1881

Days Inn Marketplace, 612 E. Marshall St., 649-2378

Embassy Suites Hotel, 2925 Emerywood Pkwy., 672-8585

Holiday Inn Downtown, 301 W. Franklin St., 644-9871

Hyatt Richmond, 6624 W. Broad St., 285-1234

Jefferson Sheraton Hotel, Franklin & Adams sts., 788-8000

Omni Richmond Hotel, 100 S. 12th St., 344-7000

Radisson Hotel, 555 East Canal St., 788-0900

Richmond Marriott, 500 E. Broad St., 643-3400

Sheraton Airport Inn, 4700 S. Laburnum Ave., 226-4300

Sheraton Park South, 9901 Midlothian Turnpike, 323-1144

SELECTED RESTAURANTS:

The Butlery, 6221 River Rd., 282-9711

Byram's Lobster House, 3215 W. Broad St., 355-9193

Ellingtons, in the Embassy Suites Hotel, 672-8585

Gallego, in the Omni Richmond Hotel, 344-7000

Hugo's, in the Hyatt Richmond, 285-1234

Julian's Restaurant, 2617 W. Broad St., 359-0605

Kabuto—Japanese House of Steaks, 8052 W. Broad St., 747-9573

Little Italy, 4350 S. Laburnum Ave., 222-7877

Sal Federico's, 1808 Staples Mill Rd., 358-9111

Traveller's Restaurant, 707 E. Franklin St., 644-1040

SELECTED ATTRACTIONS:

Berkeley Plantation, VA Hwy. 5, Charles City County, 829-6018

Edgar Allan Poe Museum, 1914 E. Main St., 648-5523

Hollywood Cemetery, Albemarle & Cherry sts., 648-8501

Maymont House and Park (turn-of-the-century estate), 1700 Hampton St., 358-7166

Museum and White House of the Confederacy, 12th & Clay sts., 649-1861

Richmond National Battlefield Park, 3215 E. Broad St., 226-1981

St. John's Church ("Give me Liberty or give me death" speech site), 24th & E. Broad sts., 648-5015

Shockoe Slip (historic district), Cary St. between 12th & 14th sts.

Virginia Aviation Museum, Huntsman Rd. next to Richmond International Airport, 371-0371

Virginia Museum of Fine Arts, 2800 Grove Ave., 367-0844

INFORMATION SOURCES:

Metropolitan Richmond Convention & Visitors Bureau
300 E. Main St.
Richmond, Virginia 23219
(804) 782-2777

Metro Richmond Visitors Center
  1710 Robin Hood Rd.
  Richmond, Virginia 23220
  (804) 358-5511
Metro Richmond Chamber of Commerce
  201 E. Franklin
  Richmond, Virginia 23219
  (804) 648-1234

Rochester, New York

Population: (*1,002,410)
  231,636 (1990C)
Altitude: 515 feet
Average Temp.: Jan., 24°F.; July, 71°F.
Telephone Area Code: 716
Time: 974-1616   **Weather:** 974-1212
Time Zone: Eastern

AIRPORT TRANSPORTATION:

Six miles to downtown Rochester.
Taxicab, major hotel courtesy car and city
  bus service.

SELECTED HOTELS:

Genesee Plaza Holiday Inn, 120 Main St.
  East, 546-6400
Holiday Inn—Airport, 911 Brooks Ave.,
  328-6000
Hyatt Regency Rochester, 125 E. Main St.,
  546-1234
Radisson Inn, 175 Jefferson Rd., 475-1910
Ramada Inn, 1273 Chili Ave., 464-8800
Rochester Airport—Howard Johnson,
  1100 Brooks Ave., 235-6030
Rochester Marriott Airport, 1890 W.
  Ridge Rd., 225-6880
Rochester Marriott Thruway Hotel, 5257
  W. Henrietta Rd., 359-1800
Stouffer Rochester Plaza Hotel, 70 State
  St., 546-3450
Strathallan Hotel, 550 East Ave., 461-5010

SELECTED RESTAURANTS:

Chapels, 30 W. Broad St., 232-2300
Daisy Flour Mill, 1880 Blossom Rd.,
  381-1880
Depot Restaurant, 41 N. Main St.,
  Pittsford, 381-9991
Edward's, 13 S. Fitzhugh, 423-0140
Jim Rund's Seafood & Steakhouse, 2851
  W. Henrietta Rd., 424-2424
Lloyd's, 289 Alexander St., 546-2211
Maplewood Inn, 3500 East Ave., 381-7700
Richardson's Canal House, 1474 Marsh
  Rd., Pittsford, 248-5000
Sabrinas, in the Strathallan Hotel,
  461-5010
Spring House, 3001 Monroe Ave.,
  586-2300

SELECTED ATTRACTIONS:

Casa Larga Vineyards (tours), 2287 Turk
  Hill Rd., Fairport, 223-4210
Genesee Country Village, Flint Hill Rd.,
  Mumford, 538-6822
Historic Brown's Race District, Platt St. &
  Brown's Race, 546-3070
International Museum of Photography &
  George Eastman House, 900 East Ave.,
  271-3361
Memorial Art Gallery of the University of
  Rochester, 500 University Ave.,
  473-7720
Rochester Museum & Science Center,
  657 East Ave., 271-4320

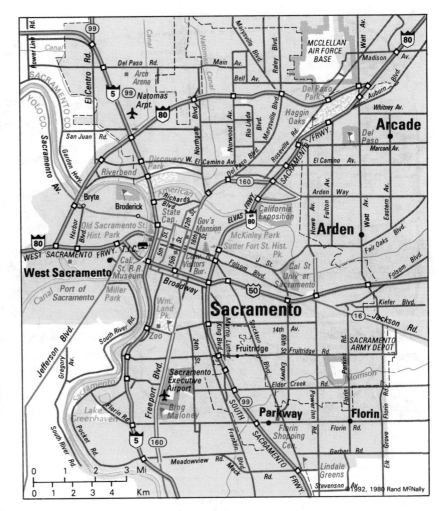

Rochester Philharmonic Orchestra, Gibbs
  & Main, 454-2620
Seabreeze Amusement Park, 4600 Culver
  Rd., 323-1900
The Strong Museum, 1 Manhattan
  Square, 263-2700
Upper Falls of the Genesee, Pont
  deRennes bridge over the Genesee
  River

INFORMATION SOURCES:

Greater Rochester Visitors Association
  126 Andrews St.
  Rochester, New York 14604-1102
  (716) 546-3070
Greater Rochester Metro Chamber of
Commerce
  55 Saint Paul St.
  Rochester, New York 14604
  (716) 454-2220

Sacramento, California

Population: (*1,481,102)
  369,365 (1990C)
Altitude: 25 feet
Average Temp.: Jan., 45°F.; July, 75°F.
Telephone Area Code: 916
Time: 767-8900   **Weather:** 442-1468
Time Zone: Pacific

AIRPORT TRANSPORTATION:

Twelve miles from Sacramento
  Metropolitan to downtown
  Sacramento.
Taxicab, airporter limousine and hotel
  limousine service.

SELECTED HOTELS:

Best Western Sandman, 236 Jibboom St.,
  443-6515
Beverly Garland Hotel, 1780 Tribute Rd.,
  929-7900
Canterbury Inn, 1900 Canterbury Rd.,
  927-3492
Capitol Plaza Holiday Inn, 300 J St.,
  446-0100
Clarion Hotel, 700 16th St., 444-8000
Fountain Suites, 321 Bercut Dr., 441-1444
Host Hotel, 6945 Airport Blvd., 922-8071
Hotel El Rancho Resort & Conference
  Center, 1029 W. Capitol Ave., West
  Sacramento, 371-6731
Hyatt Regency Hotel, 1209 L St., 443-1234
Radisson Hotel, Hwy. 160 and Canterbury
  Rd., 922-2020
Red Lion Hotel, 2001 Point West Way,
  929-8855
Red Lion Sacramento Inn, 1401 Arden
  Way, 922-8041
Sacramento Hilton, 2200 Harvard,
  922-4700

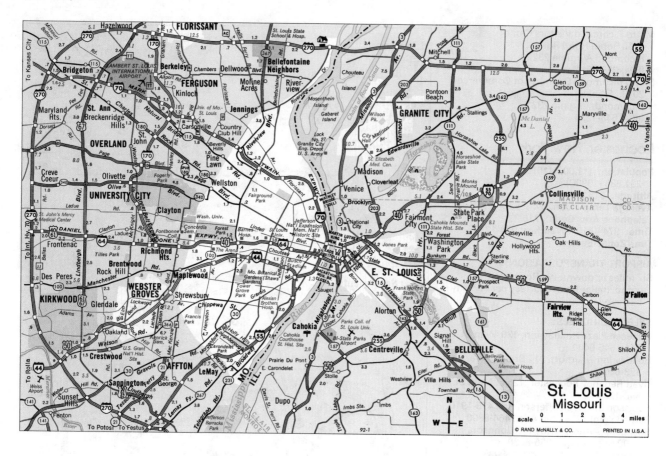

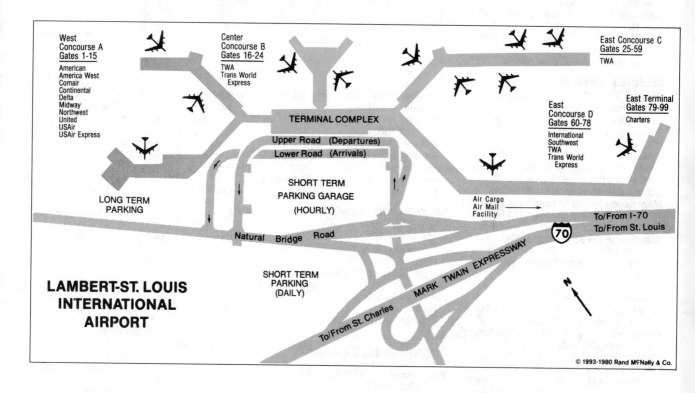

Sheraton Sunrise Hotel & Towers, 11211
Point East Dr., Rancho Cordova,
638-1100

SELECTED RESTAURANTS:
A Shot of Class, 1020 11th St., 447-5340
Aldo's Restaurant, 2914 Pasatiempo Ln.,
483-5031
Biba, 2801 Capitol Ave., 455-2422
California Fat's, 1015 Front St., 441-7966
The Firehouse, 1112 2nd St., 442-4772
Frank Fat's, 806 L St., 442-7092
Jeremiah's, 4241 Florin Rd., 392-4300
John Q's Penthouse, Holiday Inn
Downtown, 300 J St., 446-0100
Mitchell's, 544 Pavillion Ln., 488-7285
Pheasant Club, 2525 Jefferson Blvd.,
371-9530

SELECTED ATTRACTIONS:
Blue Diamond Growers (almond
industry), 446-8409
California State Historic Railroad
Museum, 125 I St., 448-4466
Crocker Art Museum, 216 O St., 264-5423
Historic Governor's Mansion, 1526 H St.,
323-3047
Sacramento History Center, 101 I St., Old
Sacramento, 264-7057
Sacramento Science Center, 3615 Auburn
Blvd., 277-6180
Sacramento Zoo, in William Land Park,
corner of Sutterville Rd. & Land Park
Dr., 264-5885
State Capitol, 10th St. & Capitol Mall,
324-0333
Sutter's Fort, 27th & L sts., 445-4422
Waterworld USA, 1600 Exposition Blvd.,
924-0555

INFORMATION SOURCES:
Sacramento Convention & Visitors
Bureau
1421 K St.
Sacramento, California 95814
(916) 264-7777
Sacramento Metropolitan Chamber of
Commerce
917 7th St.
Sacramento, California 95814
(916) 552-6800

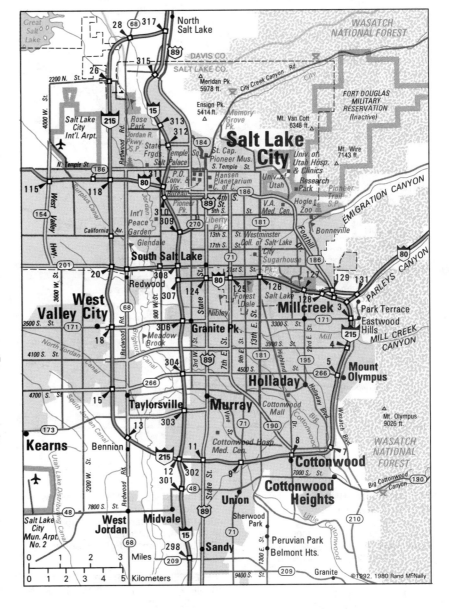

St. Louis, Missouri

Population: (*2,444,099)
396,685 (1990C)
Altitude: 470 feet
Average Temp.: Jan., 32°F.; July, 79°F.
Telephone Area Code: 314
Time: 321-2522 **Weather:** 321-2222
Time Zone: Central

AIRPORT TRANSPORTATION:
Fifteen miles to downtown St. Louis.
Taxicab and limousine bus service.

SELECTED HOTELS:
Adam's Mark, 4th & Chestnut, 241-7400
Cheshire Inn & Lodge, 6300 Clayton Rd.,
647-7300
Clarion Hotel, 200 S. 4th St., 241-9500
The Frontenac Grand Hotel, 1335 S.
Lindbergh Blvd., 993-1100
Hotel Majestic, 1019 Pine St., 436-2355
Hyatt Regency, 1 St. Louis Union Station,
231-1234

Marriott's Pavilion Hotel, 1 S. Broadway,
421-1776
Ritz Carlton St. Louis, 100 Carondelet
Plaza, 863-6300
St. Louis Airport Hilton, 10330 Natural
Bridge Rd., 426-5500
St. Louis Airport Marriott, I-70 at Lambert
Int'l Airport, 423-9700
Seven Gables Inn, 26 N. Meramec,
863-8400
Stouffer Concourse Hotel, 9801 Natural
Bridge Rd., 429-1100

SELECTED RESTAURANTS:
Al Baker's, 8101 Clayton Rd., 863-8878
Catfish & Crystal, 409 N. 11th St.,
231-7703
Dierdorf & Hart's, at St. Louis Union
Station, 421-1772
Dominic's, 5101 Wilson Ave., 771-1632
Giovanni's on the Hill, 5201 Shaw,
772-5958
Henry VIII, in the Henry VIII Hotel, 4690 N.
Lindbergh Blvd., Bridgeton, 731-3040

Kemoll's Restaurant, 211 N. Broadway,
421-0555
Tony's, 826 N. Broadway St., 231-7007

SELECTED ATTRACTIONS:
Anheuser-Busch Brewery, 12th & Lynch,
577-2626
Gateway Arch, 425-4465
Gateway Riverboat Cruises, 621-4040
Grant's Farm, 10501 Gravois, 843-1700
Missouri Botanical Garden, 4344 Shaw,
577-5100
St. Louis Art Museum, One Fine Arts Dr.,
Forest Park, 721-0067
St. Louis Science Center, 5050 Oakland
Ave., Forest Park, 289-4400
St. Louis Union Station, Market St.
between 18th & 20th, 421-6655
St. Louis Zoo, US 40 & Hampton Ave.,
Forest Park, 781-0900
Six Flags Over Mid-America, Eureka,
938-5300

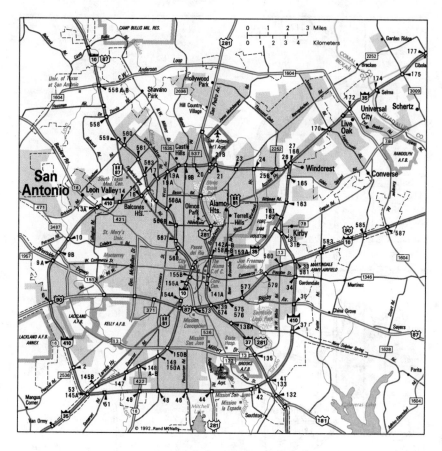

Lagoon Amusement Park & Pioneer
   Village, 17 mi. north on I-15, 451-0101
Museum of Church History & Art, 45 N.
   West Temple, 240-3310
Pioneer Trail State Park, 2601 E.
   Sunnyside Ave., 584-8392
Raging Waters, 1200 West 1700 South,
   973-4020
Temple Square, 50 W. North Temple,
   240-2534
Utah Museum of Fine Arts, AAC 101,
   Univ. of Utah campus, 581-7049
Utah State Capitol, 350 N. Main, 538-3000
Violin Making School of America (tours
   by appointment), 308 East 200 South,
   364-3651

INFORMATION SOURCES:
Salt Lake Convention and Visitors Bureau
   180 S. West Temple St.
   Salt Lake City, Utah 84101
   (801) 521-2822
Salt Lake Area Chamber of Commerce
   175 E. 400 S., Suite 600
   Salt Lake City, Utah 84111
   (801) 364-3631

San Antonio, Texas

Population: (*1,302,099)
   935,933 (1990C)
Altitude: 505 to 1,000 feet
Average Temp.: Jan., 51°F.; July, 84°F.
Telephone Area Code: 210
Time: 226-3232 **Weather:** 828-0683
Time Zone: Central

AIRPORT TRANSPORTATION:
Eight miles to downtown San Antonio.
Taxicab and limousine bus service.

SELECTED HOTELS:
Best Western Continental Inn, 9735 I-35
   N., 655-3510
Embassy Suites Northwest, 7750
   Briaridge, 340-5421
Hilton Palacio del Rio, 200 S. Alamo St.,
   222-1400
Hyatt Regency San Antonio, 123 Losoya,
   222-1234
La Mansion del Rio, 112 College St.,
   225-2581
Plaza San Antonio Hotel, 555 S. Alamo
   St., 229-1000
St. Anthony Hotel, 300 E. Travis St.,
   227-4392
San Antonio Marriott on the Riverwalk,
   711 E. Riverwalk, 224-4555
The Sheraton Fiesta, 37 NE. Loop 410,
   366-2424

SELECTED RESTAURANTS:
Chez Ardid Restaurant Gastronomique,
   1919 San Pedro, 732-3203
Crystal Baking Co., 1039 NE. Loop 410,
   826-2371
Fig Tree, 515 Paseo de la Villita, 224-1976
Grey Moss Inn, 19010 Scenic Loop Rd.,
   695-8301
Las Canarias, in the La Mansion del Rio,
   225-2581
Paesano's, 1715 McCullough Ave.,
   226-9541
Restaurant Biga, 206 E. Locust, 225-0722

INFORMATION SOURCES:
St. Louis Convention & Visitors
Commission
   10 S. Broadway, Suite 1000
   St. Louis, Missouri 63102
   (314) 421-1023; (800) 247-9791
St. Louis Regional Commerce & Growth
Association
   100 S. 4th St., Suite 500
   St. Louis, Missouri 63102
   (314) 231-5555

Salt Lake City, Utah

Population: (*1,072,227)
   159,936 (1990C)
Altitude: 4,260 feet
Average Temp.: Jan., 27°F.; July, 77°F.
Telephone Area Code: 801
Time: 467-8463 **Weather:** 575-7669
Time Zone: Mountain

AIRPORT TRANSPORTATION:
Six miles to downtown Salt Lake City.
Taxicab, bus, and limousine bus service.

SELECTED HOTELS:
Airport Hilton, 5151 Wiley Post Way,
   539-1515
Doubletree Hotel, 215 W. South Temple,
   531-7500
Holiday Inn—Downtown, 230 W. 600
   South, 532-7000
Little America Hotel, 500 S. Main St.,
   363-6781

Red Lion Salt Lake, 255 S. West Temple,
   328-2000
Salt Lake City Marriott Hotel, 75 S. West
   Temple, 531-0800
Salt Lake Hilton, 150 W. 500 South,
   532-3344

SELECTED RESTAURANTS:
Benihana of Tokyo, 165 S. West Temple
   St., 322-2421
Cafe Pierpont, 122 W. Pierpont Ave.,
   364-1222
Cowboy Grub, 2350½ Foothill Blvd.,
   466-8334
La Caille at Quail Run, 9565 Wasatch
   Blvd., 942-1751
La Fleur de Lys, 39 Post Office Pl.,
   359-5753
Market Street Grill, 54 Post Office Pl.,
   322-4668
Mikado Japanese Restaurant, 67 W. 100
   South St., 328-0929
The New Yorker, 60 Post Office Pl.,
   363-0166
Ristorante Della Fontana, 336 S. 4th East,
   328-4243

SELECTED ATTRACTIONS:
Beehive House, 67 E. South Temple,
   240-2671
Family History Library, 35 N. West
   Temple, 240-2331
Hansen Planetarium, 15 S. State St.,
   538-2098
Hogle Zoo, 2600 E. Sunnyside Ave.,
   582-1631

SELECTED ATTRACTIONS:

The Alamo, Alamo Plaza, 225-1391
Brackenridge Park, 3500 block of St.
  Mary's St., 299-8480
La Villita Historical District, on the River
  Walk between S. Alamo & E. Nueva
  sts., 299-8610
Market Square (El Mercado), 514 W.
  Commerce St., 299-8600
The River Walk, downtown San Antonio
San Antonio Missions National Historical
  Park, 2202 Roosevelt Ave., 229-5701
San Antonio Zoo & Aquarium, 3903 N. St.
  Mary's St., 734-7183
Tower of the Americas, in HemisFair
  Park, 299-8615
University of Texas Institute of Texan
  Cultures at San Antonio, 801 S. Bowie
  St., 226-7651
Witte Museum, 3801 Broadway St.,
  820-2169

INFORMATION SOURCES:

San Antonio Convention and Visitors
Bureau
  P.O. Box 2277
  San Antonio, Texas 78298
  (210) 270-8700; (800) 447-3372
Visitor Information Center
  317 Alamo Plaza
  San Antonio, Texas 78205
  (210) 299-8155
Chamber of Commerce of Greater San
Antonio
  602 E. Commerce St.
  San Antonio, Texas 78205
  (210) 229-2100

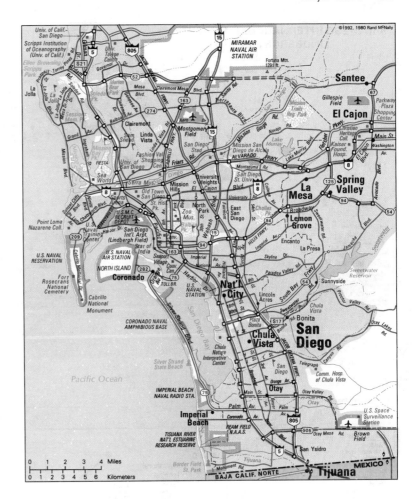

San Diego, California

Population: (*2,498,016)
  1,110,549 (1990C)
Altitude: Sea level to 823 feet
Average Temp.: Jan., 57°F.; July, 71°F.
Telephone Area Code: 619
Time: 853-1212 **Weather:** 289-1212
Time Zone: Pacific

AIRPORT TRANSPORTATION:

Three miles to downtown San Diego.
Taxicab and bus service.

SELECTED HOTELS:

Bristol Court, 1055 First Ave., 232-6141
Glorietta Bay Inn, 1630 Glorietta Blvd.,
  435-3101
Hanalei Hotel, 2270 Hotel Circle N.,
  297-1101
Holiday Inn Embarcadero, 1355 N. Harbor
  Dr., 232-3861
Hotel del Coronado, 1500 Orange Ave.,
  Coronado, 435-6611
Hyatt Islandia, 1441 Quivira Rd., 224-1234
Mission Valley Inn, 875 Hotel Circle S.,
  298-8281
Rancho Bernardo Inn, 17550 Bernardo
  Oaks Dr., 487-1611
San Diego Hilton, 1775 E. Mission Bay
  Dr., 276-4010
Sheraton Grand, 1590 Harbor Island Dr.,
  291-6400
Sheraton Harbor Island Hotel—East, 1380
  Harbor Island Dr., 291-2900
Town & Country Hotel, 500 Hotel Circle
  N., 291-7131
Westgate Hotel, 1055 2nd Ave., 238-1818

SELECTED RESTAURANTS:

Anthony's Star of the Sea Room, Harbor
  Dr. & Ash, 232-7408
El Bizcocho, in the Rancho Bernardo Inn,
  487-1611
Gourmet Room, in the Town & Country
  Hotel, 291-7131
The Marine Room, 2000 Spindrift Dr., La
  Jolla, 459-7222
Mister A's, 2550 5th Ave., Financial
  Center, 239-1377
Old Trieste, 2335 Morena Blvd., 276-1841
Thee Bungalow, 4996 W. Point Loma
  Blvd., 224-2884
Tom Ham's Lighthouse, 2150 Harbor
  Island Dr., 291-9110
Top O' the Cove, 1216 Prospect, La Jolla,
  454-7779

SELECTED ATTRACTIONS:

Balboa Park (museums, theaters,
  gardens, galleries), Center of city
Cabrillo National Monument, on the tip of
  Point Loma, 557-5450
Old Town State Historic Park, (Visitor
  Center) 4002 Wallace St., 237-6770
San Diego Zoo, off Park Blvd. in Balboa
  Park, 234-3153
Sea World of California, on Sea World
  Dr., 226-3901
Wild Animal Park, 15500 San Pasqual
  Valley Rd., Escondido, 234-6541

INFORMATION SOURCES:

San Diego Convention and Visitors
Bureau
  1200 Third Ave., Suite 824
  San Diego, California 92101
  (619) 232-3101
International Visitor Information Center
  11 Horton Plaza
  1st Ave. & F St.
  San Diego, California 92101
  (619) 236-1212
Greater San Diego Chamber of
Commerce
  402 W. Broadway, Suite 1000
  San Diego, California 92101
  (619) 232-0124

San Francisco-Oakland, California

Population: San Francisco (*1,603,678)
  723,959; Oakland (*2,082,914) 372,242
  (1990C)
Altitude: Sea level to 934 feet
Average Temp.: San Francisco Jan.,
  50°F.; July, 59°F. Oakland Jan., 48°F.;
  July, 63°F.
Telephone Area Code: (San Francisco)
  415; (Oakland) 510
Time: 767-8900 **Weather:** San
  Francisco 936-1212; Oakland 936-1212
Time Zone: Pacific

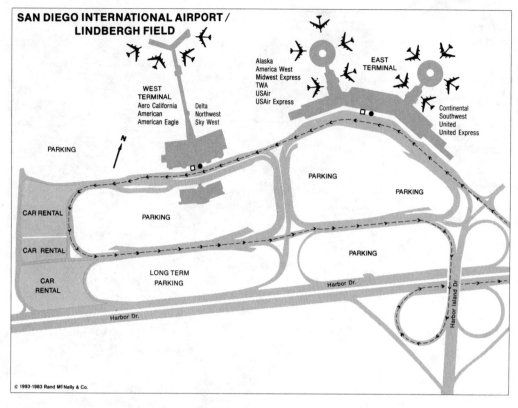

SAN DIEGO INTERNATIONAL AIRPORT / LINDBERGH FIELD

WEST TERMINAL
Aero California
American
American Eagle

Delta
Northwest
Sky West

Alaska
America West
Midwest Express
TWA
USAir
USAir Express

EAST TERMINAL

Continental
Southwest
United
United Express

PARKING

CAR RENTAL

CAR RENTAL

CAR RENTAL

PARKING

LONG TERM PARKING

PARKING

PARKING

PARKING

Harbor Dr.

Harbor Dr.

Harbor Island Dr.

© 1993-1983 Rand M°Nally & Co.

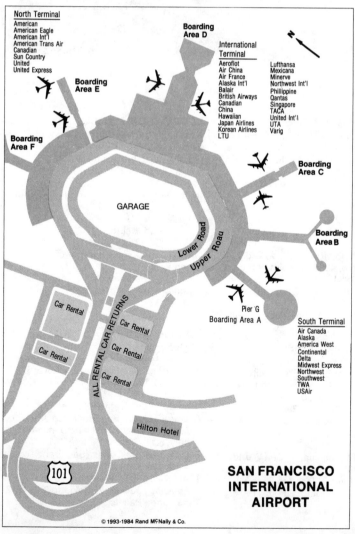

North Terminal
American
American Eagle
American Int'l
American Trans Air
Canadian
Sun Country
United
United Express

Boarding Area D

International Terminal
Aeroflot
Air China
Air France
Alaska Int'l
Balair
British Airways
Canadian
China
Hawaiian
Japan Airlines
Korean Airlines
LTU

Lufthansa
Mexicana
Minerve
Northwest Int'l
Phillippine
Qantas
Singapore
TACA
United Int'l
UTA
Varig

Boarding Area E

Boarding Area F

GARAGE

Boarding Area C

Lower Road

Upper Road

Boarding Area B

Car Rental

ALL RENTAL CAR RETURNS

Car Rental

Car Rental

Car Rental

Car Rental

Pier G
Boarding Area A

South Terminal
Air Canada
Alaska
America West
Continental
Delta
Midwest Express
Northwest
Southwest
TWA
USAir

Hilton Hotel

101

SAN FRANCISCO INTERNATIONAL AIRPORT

© 1993-1984 Rand M°Nally & Co.

AIRPORT TRANSPORTATION:
Fifteen miles to downtown San Francisco from San Francisco International; 25 miles to downtown Oakland from San Francisco International Airport. Eight miles to downtown Oakland from Oakland International.

Taxicab, bus, and limousine bus service to downtown San Francisco. Taxicab, limousine, bus, and BART service to downtown Oakland.

SELECTED HOTELS: SAN FRANCISCO
The Donatello, 501 Post St., 441-7100
The Fairmont Hotel, 950 Mason St., 772-5000
Four Seasons-Clift, 495 Geary St., 775-4700
Hilton Square, O'Farrell St. between Taylor & Mason, 771-1400
The Holiday Inn Union Square, 480 Sutter St., 398-8900
Huntington Hotel, 1075 California St., 474-5400
Hyatt on Union Square, 345 Stockton St., 398-1234
Hyatt Regency San Francisco, 5 Embarcadero Center, 788-1234
Mark Hopkins Inter-Continental, 999 California St., 392-3434
Miyako Hotel, 1625 Post St., 922-3200
The Phoenix Hotel, 601 Eddy St., 776-1380
Queen Anne, 1590 Sutter St., 441-2828
The Stouffer Stanford Court Hotel, 905 California St., 989-3500
The Westin St. Francis, 335 Powell St., 397-7000

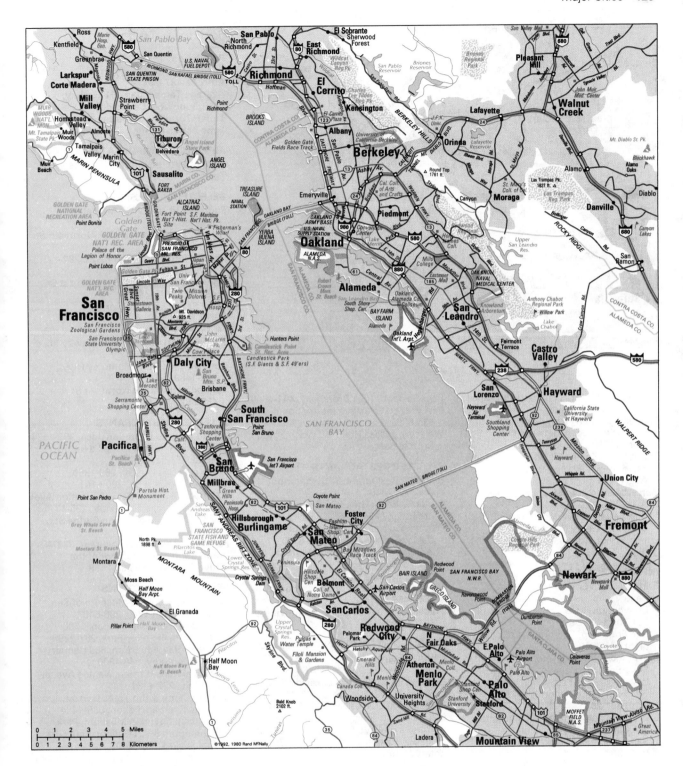

SELECTED RESTAURANTS: SAN FRANCISCO

Amelio's, 1630 Powell St., North Beach, 397-4339

Blue Fox Restaurant, 659 Merchant St., 981-1177

Empress of China, 838 Grant Ave., 434-1345

Ernie's Restaurant, 847 Montgomery St., 397-5969

Fleur de Lys, 777 Sutter St., 673-7779

Fournou's Ovens, in The Stanford Court Hotel, 989-1910

Tommy Toy's Haute Cuisine Chinoise, 655 Montgomery St., 397-4888

The Waterfront Restaurant, Pier 7, The Embarcadero, 391-2696

SELECTED ATTRACTIONS: SAN FRANCISCO

Alcatraz Island, San Francisco Bay, 546-2805

Chinatown, Grant Ave. & Stockton St. (begins at Grant & Bush)

Coit Tower/Telegraph Hill, northeast San Francisco, reached via Lombard St.

Exploratorium, 3601 Lyon St., 561-0360

Fisherman's Wharf, on Jefferson St. between Hyde & Taylor sts.

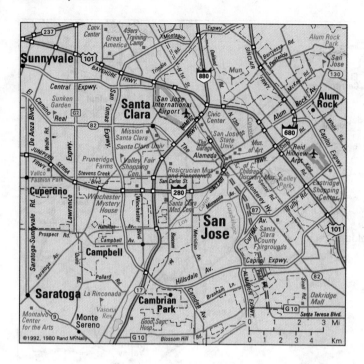

San Jose, California

Population: (*1,497,577)
782,248 (1990C)
Altitude: 87 feet
Average Temp.: Jan., 51°F.; July, 65°F.
Telephone Area Code: 408
Time: 767-8900   **Weather:** (none)
Time Zone: Pacific

AIRPORT TRANSPORTATION:
Five miles to downtown San Jose.
Taxicab and bus service.

SELECTED HOTELS:
Fairmont Hotel, 170 S. Market St.,
998-1900
Holiday Inn—Airport, 1355 North 4th St.,
453-5340
Holiday Inn Park Center Plaza, 282
Almaden Blvd., 998-0400
Howard Johnson's, 1755 North 1st St.,
453-3133
Hyatt San Jose, 1740 North 1st St.,
993-1234
Le Baron, 1350 North 1st St., 453-6200
The Red Lion Hotel, 2050 Gateway Pl.,
453-4000
Rodeway Inn, 2112 Monterey Hwy.,
294-1480
Vagabond Inn, 1488 North 1st St.,
453-8822

SELECTED RESTAURANTS:
Amalfi's, in the Hyatt San Jose, 993-1234
Belvedere Gardens, in the Le Baron
Hotel, 453-6200
Emile's, 545 S. 2nd St., 289-1960
Eulipia, 374 S. 1st St., 280-6161
L'Horizon Restaurant, 1250 Aviation Ave.,
at the San Jose Jet Center, 295-1771
Lou's Village, 1465 W. San Carlos St.,
293-4570
Paolo's Continental Restaurant, 333 W.
San Carlos, ground floor of the
Riverfront Tower, 294-2558

SELECTED ATTRACTIONS:
Children's Discovery Museum of San
Jose, 180 Woz Way, 298-5437
Japanese Friendship Garden, 1500
Senter Rd. in Kelley Park, 287-2290
Monterey Bay Aquarium, 886 Cannery
Row, Monterey, 648-4800
Municipal Rose Garden, Naglee Ave. at
Dana St., 277-4191
Roaring Camp & Big Trees Narrow
Gauge Railroad, Felton, 335-4484
Rosicrucian Egyptian Museum, Park,
Planetarium & Art Gallery, Park &
Naglee aves., 287-2807
San Jose Historical Museum (25 acres of
original and replica 1890s buildings),
1600 Senter Rd. in Kelley Park,
287-2290
TECH Museum of Innovation, 145 W. San
Carlos St., 279-7156
Trinity Cathedral (1863), 81 N. 2nd St.,
(tours by appointment) 293-7953
Winchester Mystery House, 525 S.
Winchester Blvd., 247-2000

INFORMATION SOURCES:
San Jose Convention & Visitors Bureau
333 W. San Carlos, Suite 1000
San Jose, California 95110
(408) 295-9600; FYI Events
Line (408) 295-2265

Golden Gate Bridge, spans the entrance
to San Francisco Bay
Golden Gate Park, from Stanyan St. to
the Pacific Ocean, between Fulton St. &
Lincoln Blvd.
North Beach (night-life area), northeast
sector of San Francisco, Columbus
Ave. is the main artery
San Francisco Zoo, Sloat Blvd. & 45th
Ave., 753-7083
Union Square, bounded by Powell,
Stockton, Geary, & Post sts.
Union Street, from the 1600 to the 2200
block

INFORMATION SOURCES:
San Francisco Convention & Visitors
Bureau
Convention Plaza
201 Third St., Suite 900
San Francisco, California 94103
(415) 974-6900
San Francisco Visitor Information Center
Swig Pavilion, Hallidie Plaza
Powell & Market sts.
San Francisco, California 94103
(415) 391-2000
San Francisco Chamber of Commerce
465 California St., 9th Floor
San Francisco, California 94104
(415) 392-4511

SELECTED HOTELS: OAKLAND
Best Western Thunderbird Inn, 233
Broadway, 452-4565
Claremont Resort & Spa, Domingo &
Ashby aves., 843-3000
Clarion at Oakland International, 455
Hegenberger Rd., 562-6100
Holiday Inn—Bay Bridge, 1800 Powell
St., Emeryville, 658-9300
Holiday Inn Oakland Airport, 500
Hegenberger Rd., 562-5311
Oakland Airport Hilton, 1 Hegenberger
Rd., 635-5000

Parc Oakland, 1001 Broadway, 451-4000
Park Plaza, at Oakland International,
635-5300
Waterfront Plaza Hotel, 10 Washington
St., 836-3800

SELECTED RESTAURANTS: OAKLAND
Bay Wolf Restaurant, 3853 Piedmont
Ave., 655-6004
Crogan's Bar & Grill, City Square at
Broadway, 464-3698
Ducks & Company, in the Clarion at
Oakland International, 562-6100
The Gingerbread House, 741 5th St.,
444-7373
La Brasserie, 542 Grand Ave., 893-6206
Scott's Seafood Grill & Bar, 2 Broadway,
444-3456
Trader Vic's, 9 Anchor Dr., Emeryville
653-3400

SELECTED ATTRACTIONS: OAKLAND
Camron-Stanford House (1876; tours),
1418 Lakeside Dr., 836-1976
Children's Fairyland, in Lakeside Park at
Grand & Bellevue aves., 452-2259
Lakeside Park/Lake Merritt, 238-FUNN
Marine World Africa U.S.A., Marine
World Pkwy., Vallejo, (707) 643-ORCA
Oakland Museum, 1000 Oak St., 238-3514
Rotary Natural Science Center, in
Lakeside Park at Bellevue & Perkins
aves., 238-3739
Takara Sake Company, 708 Addison St.,
Berkeley, 540-8250
Trial & Show Gardens, in Lakeside Park,
238-3208

INFORMATION SOURCES: OAKLAND
Oakland Convention and Visitors Bureau
1000 Broadway, Suite 200
Oakland, California 94607-4020
(510) 839-9000
Oakland Chamber of Commerce
475 14th St.
Oakland, California 94612-1928
(510) 874-4800

San Jose Chamber of Commerce
180 S. Market St.
San Jose, California 95113
(408) 998-7000

Seattle, Washington

Population: (*1,972,961)
516,259 (1990C)
Altitude: Sea level to 520 feet
Average Temp.: Jan., 41°F.; July, 66°F.
Telephone Area Code: 206
Time: 361-8463 **Weather:** 526-6087
Time Zone: Pacific

AIRPORT TRANSPORTATION:

Fourteen miles to downtown Seattle.
Taxicab, limousine bus, Metro Transit,
and Airporter service.

SELECTED HOTELS:

Doubletree Inn, 205 Strander Blvd.,
246-8220
Four Seasons Olympic, 411 University
St., 621-1700
Holiday Inn Crowne Plaza, 6th Ave. and
Seneca St., 464-1980
Hyatt Regency—Bellevue, Bellevue Way
& NE 8th St., Bellevue, 462-1234
Radisson Seattle Airport, 17001
International Blvd., 244-6000
Red Lion Inn/Sea-Tac, 18740 Pacific Hwy.
S., 246-8600
Sea-Tac Holiday Inn Airport, 17338
Pacific Hwy. S., 248-1000
Seattle Hilton, 6th & University, 624-0500
Seattle Marriott Sea-Tac Airport, 3201 S.
176th St., 241-2000
Seattle Sheraton Hotel & Towers, 1400
6th Ave., 621-9000
Stouffer Madison, 515 Madison, 583-0300
Westin Hotel Seattle, 1900 5th Ave.,
728-1000

SELECTED RESTAURANTS:

Carvery, Seattle-Tacoma International
Airport, 433-5622
Fuller's, in the Seattle Sheraton Hotel &
Towers, 621-9000
The Hunt Club, in the Sorrento Hotel, 900
Madison, 622-6400
Jonah's, in the Bellevue Holiday Inn,
11211 Main St., Bellevue, 455-5240
Le Tastevin, 19 W. Harrison, 283-0991
Nikko, in the Westin Hotel Seattle,
728-1000
The Painted Table, in the Alexis Hotel,
1007 1st Ave. at Madison, 624-4844

SELECTED ATTRACTIONS:

Hiram M. Chittenden Locks, located
where NW. Market turns, Ballard,
783-7059
Museum of Flight, 9404 East Marginal
Way S., 764-5700
Pacific Science Center, 200 Second Ave.
N. in Seattle Center, 443-2001
Pike Place Market, First & Pike sts.,
682-7453
Seattle Aquarium, Pier 59 in Waterfront
Park, 386-4300
Seattle Art Museum, 100 University St.,
625-8900
Seattle Underground Tour, 682-1511
Space Needle, 219 Fourth Ave. N. in
Seattle Center, 443-2100

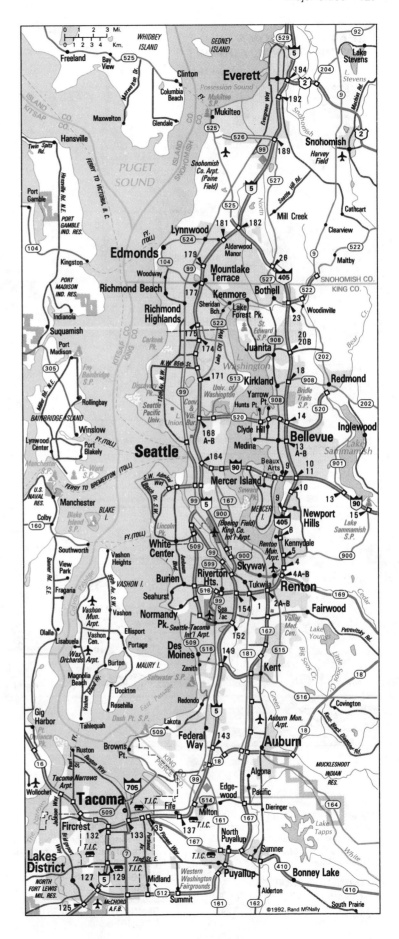

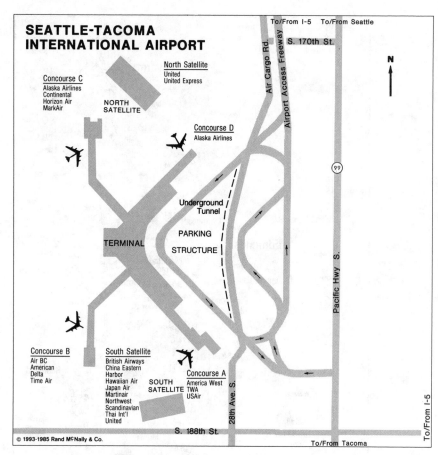

SEATTLE-TACOMA INTERNATIONAL AIRPORT

To/From I-5   To/From Seattle

S. 170th St.

Air Cargo Rd.

Airport Access Freeway

North Satellite
United
United Express

Concourse C
Alaska Airlines
Continental
Horizon Air
MarkAir

NORTH SATELLITE

Concourse D
Alaska Airlines

Underground Tunnel

TERMINAL

PARKING STRUCTURE

99

N

Pacific Hwy S.

Concourse B
Air BC
American
Delta
Time Air

South Satellite
British Airways
China Eastern
Harbor
Hawaiian Air
Japan Air
Martinair
Northwest
Scandinavian
Thai Int'l
United

SOUTH SATELLITE

Concourse A
America West
TWA
USAir

28th Ave. S.

To/From I-5

S. 188th St.

To/From Tacoma

© 1993-1985 Rand McNally & Co.

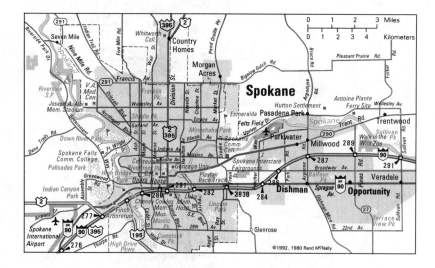

0 1 2 3 Miles
0 1 2 3 4 Kilometers

291 · Seven Mile · Whitworth Coll. · 395 · 2 · Country Homes · Pend Oreille Hy.

Indian Trail Rd. · Nine Mile Rd. · Five Mile Rd. · Wall St. · Morgan Acres · Bigelow Gulch Rd. · Pleasant Prairie Rd. · Forker Rd.

Riverside S.P. · V.A. Med Cen · Francis Av. · Division St. · Nevada St. · Crestline St. · Market St. · Argonne Rd. · **Spokane** · Antoine Plante Ferry Site · Wellesley Av.

Joseph A. Albi Mem. Stadium · Wellesley Av. · Driscoll Blvd. · Northwest Blvd. · Esmeralda · Pasadena Park · Hutton Settlement

291 · Shadle Pk. · Garland Av. · Empire Av. · Felts Field · Upriver Dr. · Spokane · Trent Rd. · Trentwood

Down River Park · Deno Trails Rd. · Ash St. · Maple St. · Indiana Av. · Illinois Av. · Minnehaha Park · 2 395 · Parkwater · 290 · Walk in the Pk Wild Zoo · Sullivan Rd.

Spokane Falls Comm. College · Palisades Park · Ft. Wright Dr. · Mission Av. · Spokane Comm. Coll. · Boone Av. · Gonzaga Univ. · 289 Millwood · Broadway Av. · 291

High Bridge Park · Greenwood · Coliseum · Riverfront Pk. · Opera House · Spokane Interstate Fairgrounds · 287 · Park Rd. · Pines Rd. · 291 · Veradale

Indian Canyon Park · 2 · 201 · 280 · 282 · 283B · 284 · Playfair Race Track · 286 · Dishman · Sprague Av. · Balfour Pk · **Opportunity** · Evergreen Rd.

Spokane International Airport · 277 · 90 90 395 · 195 · Finch Arboretum · Cheney Cowles Mem. Mus. · St. Luke's Mem. Hosp. · Lincoln Park · Manito Pk. · High Dr. · Comstock Pk. · Glenrose · 27 · Dishman Mica Rd. · 22nd Av. · Terrace View Pk. · Sullivan Rd.

276 · Thorpe Rd. · High Drive Pkwy. · Regal St.

©1992, 1980 Rand McNally

Tillicum (Indian) Village Tour, Blake
  Island State Park (tour starts from Pier
  56 on the Seattle Waterfront), 443-1244
Woodland Park Zoo, 5500 Phinney Ave.
  N., 684-4800

INFORMATION SOURCES:
Seattle-King County Convention &
Visitors Bureau/Convention Sales
  520 Pike St., Suite 1300

Seattle, Washington 98101
(206) 461-5800
Seattle-King County Convention &
Visitors Bureau/Visitor Information
  Level 1, Galleria
Washington State Convention & Trade
Center
  800 Convention Pl.
Seattle, Washington 98101
(206) 461-5840

Seattle Chamber of Commerce
  600 University St., Suite 1200
  Seattle, Washington 98101
  (206) 389-7200

Spokane, Washington

Population: (*361,364)
  177,196 (1990C)
Altitude: 1,898 feet
Average Temp.: Jan., 26°F.; July, 70°F.
Telephone Area Code: 509
Time: 484-3344 **Weather:** 624-8905
Time Zone: Pacific

AIRPORT TRANSPORTATION:
Five miles to downtown Spokane.
Taxicab, airporter limousine and
  limousine service.

SELECTED HOTELS:
Best Western Trade Winds "Downtown,"
  W. 907 3rd Ave., 838-2091
Best Western Trade Winds—North, N.
  3033 Division St., 326-5500
Cavanaugh's Inn at the Park, W. 303 N.
  River Dr., 326-8000
Cavanaugh's River Inn, N. 700 Division
  St., 326-5577.
Courtyard by Marriott, N. 401 Riverpoint
  Blvd., 456-7600
Friendship Inn, 4301 W. Sunset Hwy.,
  838-1471
Holiday Inn Spokane West, 4212 W.
  Sunset Blvd., 747-2021
Ramada Inn, at Spokane International
  Airport, 838-5211
Red Lion Motor Inn, N. 1100 Sullivan Rd.,
  924-9000
Sheraton—Spokane, N. 322 Spokane
  Falls Ct., 455-9600
Shilo Inn, E. 923 3rd Ave., 535-9000
West Coast Ridpath, 515 W. Sprague
  Ave., 838-2711

SELECTED RESTAURANTS:
Ankeny's, in the West Coast Ridpath,
  838-6311
Chapter Eleven, E. 105 Mission Ave.,
  326-0466
1881, in the Sheraton—Spokane,
  455-9600
The Mustard Seed, W. 245 Spokane Falls
  Blvd., 747-2689
The Onion Bar and Grill, W. 302
  Riverside, 747-3852
Patsy Clark's Mansion, W. 2208 2nd Ave.,
  838-8300
Sea Galley, N. 1221 Howard Ave.,
  327-3361
Spokane House, in the Friendship Inn,
  838-1475
Stockyard Inn, 3827 E. Boone Ave.,
  534-1212
Town & Country, 5615 Trent Ave.,
  534-3868

SELECTED ATTRACTIONS:
Antique 1909 Loof Carousel, on Spokane
  Falls Blvd. between Washington &
  Howard, in Riverfront Park, 625-6600
Centennial Trail, follows the Spokane
  River from downtown to the
  Washington/Idaho state line
Cheney Cowles Memorial Museum
  (Eastern Washington Historical
  Society), W. 2316 First, 456-3931

Downtown Skywalk Shopping System
Manito Park/Japanese Gardens, 18th &
  Bernard, 625-6200
Riverfront Park, 625-6200

INFORMATION SOURCES:

Spokane Regional Convention and
Visitors Bureau
  W. 926 Sprague, Suite 180
  Spokane, Washington 99204
  Visitor Info. (509) 747-3230
Spokane Area Chamber of Commerce
  W. 1020 Riverside
  Spokane, Washington 99204
  (509) 624-1393

Tampa-St. Petersburg, Florida

Population: (*2,067,959—Tampa-St.
  Petersburg-Clearwater Metro Area)
  Tampa 280,150;
  St. Petersburg 238,629 (1990C)
Altitude: (Tampa) 57 feet; (St.
  Petersburg) 44 feet
Average Temp.: Jan., 61°F.; July, 82°F.
Telephone Area Code: 813
Time: 1-976-1111   **Weather:** 645-2506
Time Zone: Eastern

AIRPORT TRANSPORTATION:

Five miles from Tampa International to
  downtown Tampa; 25 miles to
  downtown St. Petersburg. Ten miles
  from St. Petersburg/Clearwater Airport
  to St. Petersburg.
Taxicab and limousine service to Tampa
  and St. Petersburg; also bus service to
  Tampa.

SELECTED HOTELS: TAMPA

Best Western, 9331 Adamo Dr., 621-5511
Guest Quarters, 3050 N. Rocky Point Dr.
  W., 888-8800
Holiday Inn—Airport, 4500 W. Cypress
  St., 879-4800
Holiday Inn Downtown, 111 W. Fortune
  St., 223-1351
Hyatt Regency—Tampa, 211 N. Tampa
  St., 225-1234
Hyatt Regency—Westshore, 6200
  Courtney Campbell Causeway,
  874-1234
Omni Hotel—Tampa, 700 N. Westshore
  Blvd., 289-8200
Radisson Bay Harbor Inn, 7700 Courtney
  Campbell Causeway, 281-8900
Ramada Inn, 400 E. Bearss Ave., 961-1000
Rodeway Safari Inn, 4139 E. Busch Blvd.,
  988-9191
Sheraton Grand, 4860 W. Kennedy Blvd.,
  286-4400
Tampa Airport Marriott, Tampa Int'l
  Airport, 879-5151
Tampa Marriott Westshore, 1001 N.
  Westshore Blvd., 287-2555

SELECTED RESTAURANTS: TAMPA

Armani's, in the Hyatt Regency—
  Westshore, 874-1234
Bern's Steak House, 1208 S. Howard
  Ave., 251-2421
C.K.'s Revolving Rooftop Restaurant, in
  the Tampa Airport Marriott, 879-5151
The Colonnade, 3401 Bayshore Blvd.,
  839-7558
The Columbia, 2117 E. 7th Ave., 248-4961

Cypress Room, in the Saddlebrook
  Resort, 25 miles north of Tampa on
  I-75, 973-1111
J. Fitzgerald's, in the Sheraton Grand,
  286-4444
Old Spaghetti Warehouse, 1911 N. 13th
  St., in Ybor Square, 248-1720
Oyster Catchers, in the Hyatt Regency—
  Westshore, 874-1234

SELECTED ATTRACTIONS: TAMPA

Adventure Island, 4545 Bougainvillea
  Ave., 987-5660
Busch Gardens, Busch Blvd. & 40th St.,
  987-5082
Children's Museum of Tampa, 7550
  North Blvd., 935-8441
Henry B. Plant Museum, 401 W. Kennedy
  Blvd., 254-1891
Lowry Park Zoo, 7530 North Blvd.,
  935-8552
Museum of Science & Industry, 4801 E.
  Fowler Ave., 985-5531
The Shops on Harbour Island, 601 S.
  Harbour Island Blvd., 228-7807

Tampa Museum of Art, 601 Doyle Carlton
  Dr., 223-8130
Ybor City State Museum, 1818 Ninth
  Ave., 247-6323
Ybor Square, 8th Ave. & 13th St.,
  247-4497

INFORMATION SOURCES: TAMPA

Tampa/Hillsborough Convention &
Visitors Association
  111 Madison St., Suite 1010
  Tampa, Florida 33602-4706
  (813) 223-1111
Greater Tampa Chamber of Commerce
  801 E. Kennedy
  P.O. Box 420
  Tampa, Florida 33601
  (813) 228-7777

SELECTED HOTELS: ST. PETERSBURG

Days Inn Marina Beach Resort, 6800 34th
  St. S., 867-1151
The Don CeSar, 3400 Gulf Blvd., St.
  Petersburg Beach, 360-1881

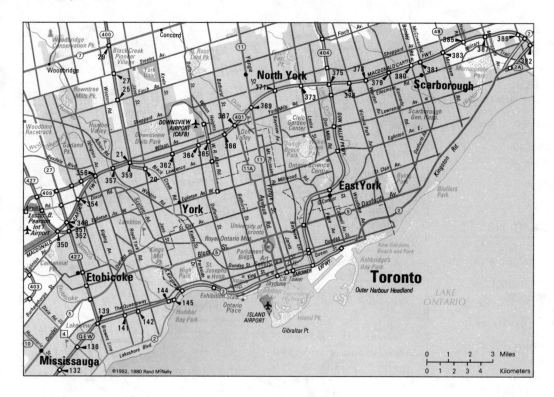

Holiday Inn—Stadium, 4601 34th St. S.,
867-3131
Lamp Post Inn, 1200 34th St. N., 323-1221
La Quinta, 4999 34th St. N., (US 19),
527-8421
The St. Petersburg Hilton and Towers,
333 First St. S., 894-5000
The Tradewinds, 5500 Gulf Blvd., St.
Petersburg Beach, 367-6461
Traveler's Inn, 2595 54th Ave. N.,
522-3191

SELECTED RESTAURANTS:
ST. PETERSBURG

Alessi's, at The Pier, 800 2nd Ave. NE,
894-4659
Billy's Tierra Verde Seafood, 1 Colony
Rd., Tierra Verde, 866-2115
The Columbia, at The Pier, 800 2nd Ave.
NE, 822-8000
The Lobster Pot, 17814 Gulf Blvd.,
Reddington Shores, 391-8592
Parker's Landing, in the Days Inn Marina
Beach Resort, 867-1151

SELECTED ATTRACTIONS:
ST. PETERSBURG

Great Explorations (hands-on museum),
1120 4th St. S., 821-8992
Museum of Fine Arts, 255 Beach Dr. NE,
896-2667
P. Buckley Moss Gallery, 190 Fourth Ave.
NE, 894-2899
The Pier, on the waterfront at the end of
2nd Ave. NE, 821-6164
St. Petersburg Historical Museum, 335
Second Ave. NE, 894-1052
Salvador Dali Museum, 1000 Third St. S.,
823-3767
Sunken Gardens, 1825 Fourth St. N.,
896-3186

INFORMATION SOURCES:
ST. PETERSBURG

Pinellas Suncoast Convention & Visitor
Bureau
  Florida Suncoast Dome
  1 Stadium Dr., Suite A
  St. Petersburg, Florida 33705
  (813) 892-7892
Suncoast Welcome Center
  2001 Ulmerton Rd.
  Clearwater, Florida 34622
  (813) 573-1449
St. Petersburg Area Chamber of
Commerce
  100 2nd Ave. N.
  P.O. Box 1371
  St. Petersburg, Florida 33731
  Visitor Info. (813) 821-4715

Toronto, Ontario, Canada

Population: (*3,427,168)
599,217 (1986C)
Altitude: 275 ft.
Average Temp.: Jan., 23°F.; July, 69°F.
Telephone Area Code: 416
Time: (none)   **Weather:** 676-3066
Time Zone: Eastern

AIRPORT TRANSPORTATION:

Eighteen miles to downtown Toronto.
Taxicab and bus service.

SELECTED HOTELS:

Four Seasons Hotel, 21 Avenue Rd.,
964-0411
Hilton International, 145 Richmond St.
W., 869-3456
Holiday Inn—Downtown, 89 Chestnut St.,
977-0707
Inn on the Lake, 1926 Lakeshore Blvd. W.,
766-4392

Inn-on-the-Park, 1100 Eglinton Ave. E.,
444-2561
King Edward Hotel, 37 King St. E.,
863-9700
Park Plaza, 4 Avenue Rd., 924-5471
Royal York, 100 Front St. W., 368-2511
Sheraton Centre, 123 Queen St., 361-1000
Sutton Place, 955 Bay St., 924-9221
Westbury Howard Johnson, 475 Yonge
St., 924-0611

SELECTED RESTAURANTS:

The Acadian Room, in the Royal York
Hotel, 368-2511
Fisherman's Wharf, 145 Adelaide St. W.,
364-1346
Lighthouse, in the Westin Harbour Castle,
1 Harbour Sq., 869-1600
Old Spaghetti Factory, 54 Esplanade,
864-9761
The Old Mill, 21 Old Mill Rd., 236-2641
The Winter Palace, in the Sheraton
Centre, 361-1000

SELECTED ATTRACTIONS:

Art Gallery of Ontario, 317 Dundas St. W.,
977-0414
Black Creek Pioneer Village, 1000 Murray
Ross Pkwy., Downsview, 736-1733
Casa Loma, 1 Austin Terrace, 923-1171
CN Tower, 301 Front St. W., 360-8500
Harbourfront, Queen's Quay, 973-3000
Metro Zoo, Meadowvale Rd. N.,
Scarborough, 392-5900
Ontario Place, 955 Lakeshore Blvd. W.,
965-7711
Ontario Science Centre, 770 Don Mills
Rd., 429-4100
Royal Ontario Museum, 100 Queen's
Park, 586-5549
SkyDome, 300 Bremner Blvd., 341-3663

INFORMATION SOURCES:

Metropolitan Toronto Convention &
Visitors Association
  207 Queens Quay W., Suite 590
  Box 126
  Toronto, Ontario M5J 1A7
  (416) 368-9990
Canadian Consulate General
  1251 Avenue of the Americas
  New York, New York 10020
  (212) 768-2400

Tucson, Arizona

Population: (*666,880)
  405,390 (1990C)
Altitude: 2,386 feet
Average Temp.: Jan., 50°F.; July, 86°F.
Telephone Area Code: 602
Time: 1-676-1676 **Weather:** 294-2522
Time Zone: Mountain Standard

AIRPORT TRANSPORTATION:

Ten miles to downtown Tucson.
Taxicab, limousine, van and bus service.

SELECTED HOTELS:

Arizona Inn, 2200 E. Elm St., 325-1541
Best Western Aztec Inn, 102 N. Alvernon
  Way, 795-0330
Best Western Executive Inn, 333 W.
  Drachman St., 791-7551
Best Western Inn at the Airport, 7060 S.
  Tucson Blvd., 746-0271
Best Western Tucson Innsuites, 6201 N.
  Oracle Rd., 297-8111
Discovery Inn, 1010 S. Freeway, 622-5871
Doubletree Hotel, 445 S. Alvernon Way,
  881-4200
Embassy Suites, 5335 E. Broadway,
  745-2700
The Embassy Suites—Airport, 7051 S.
  Tucson Blvd., 573-0700
Holiday Inn—Airport, 4550 S. Palo Verde
  Blvd., 746-1161
Lexington Hotel Suites, 7411 N. Oracle
  Rd., 575-9255
Pueblo Inn, 350 S. Freeway, 622-6611
Quality Inn University, 1601 N. Oracle
  Rd., 623-6666
Radisson Suite Hotel, 6555 E. Speedway
  Blvd., 721-7100
Ramada Inn Downtown, 475 N. Granada,
  624-8341
The Viscount Suite Hotel, 4855 E.
  Broadway Blvd., 745-6500
Wayward Winds Lodge, 707 W. Miracle
  Mile, 791-7526

SELECTED RESTAURANTS:

Charles, 6400 E. El Dorado Circle,
  296-7173
Gold Room, at the Westward Look
  Resort, 245 E. Ina Rd., 297-1151
Janos, 150 N. Main, 884-9426
Le Rendez-Vous, 3844 E. Fort Lowell,
  323-7373
Palomino, 2959 N. Swan, 325-0413
Saguaro Corners, 3750 S. Old Spanish
  Trail, 886-5424
Scordato's Restaurant, 4405 W.
  Speedway, 792-3055
The Tack Room, 2800 N. Sabino Canyon
  Rd., 722-2800

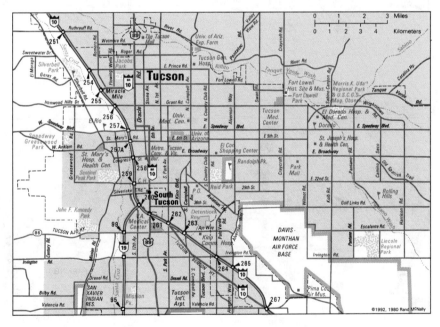

© 1993-1992 Rand McNally & Co.

©1992, 1980 Rand McNally

SELECTED ATTRACTIONS:

Arizona-Sonora Desert Museum, 2021 N.
  Kinney Rd., 883-1380
Colossal Cave, Old Spanish Trail, 22
  miles southeast of Tucson, 791-7677
Grace H. Flandrau Planetarium, at the
  University of Arizona, Cherry Ave. &
  University Blvd., 621-4515
International Wildlife Museum, 4800 W.
  Gates Pass Rd., 624-4024 or 629-0100

Old Tucson (famous movie location and
  "Old West" town), 201 S. Kinney Rd.,
  883-0100
Pima Air Museum, 6000 E. Valencia Rd.,
  574-9658
Sabino Canyon Tours, 749-2861
Tucson Botanical Gardens, 2150 N.
  Alvernon Way, 326-9255

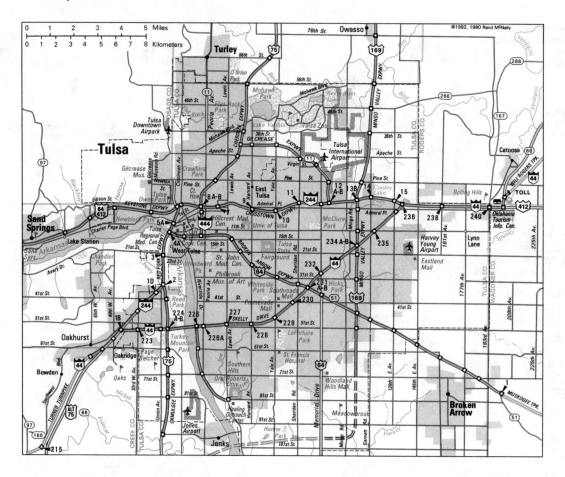

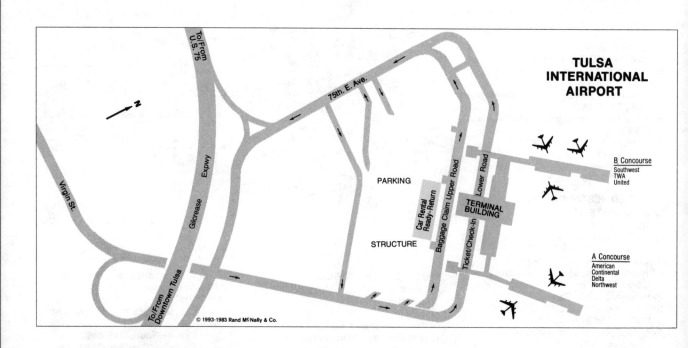

TULSA INTERNATIONAL AIRPORT

B Concourse
Southwest
TWA
United

A Concourse
American
Continental
Delta
Northwest

© 1993-1983 Rand McNally & Co.

INFORMATION SOURCES:

Metropolitan Tucson Convention and
Visitors Bureau
130 S. Scott
Tucson, Arizona 85701
(602) 624-1817
Tucson Metropolitan Chamber of
Commerce
465 W. St. Mary's Rd.
Tucson, Arizona 85702
(602) 792-1212

Tulsa, Oklahoma

Population: (*708,954)
367,302 (1990C)
Altitude: 711 feet
Average Temp.: Jan., 39°F.; July, 83°F.
Telephone Area Code: 918
Time: 477-1000 **Weather:** 743-3311
Time Zone: Central

AIRPORT TRANSPORTATION:

Eight miles to downtown Tulsa.
Taxicab and limousine service.

SELECTED HOTELS:

Best Western Trade Winds Central Inn,
3141 E. Skelly Dr., 749-5561
Best Western Trade Winds East Inn, 3337
E. Skelly Dr., 743-7931
Camelot Hotel, 4956 S. Peoria, 747-8811
Doubletree Downtown, 616 W. 7th St.,
587-8000
Holiday Inn—Tulsa Holidome, 8181 E.
Skelly Dr., 663-4541
Sheraton Inn—Tulsa Airport, 2201 N. 77
East Ave., 835-9911
Sheraton Kensington, 1902 E. 71st St.,
493-7000
S.W. Airport Hotel, 11620 E. Skelly Dr.,
437-9200
Tulsa Grand, 5000 E. Skelly Dr., 622-7000
Tulsa Marriott, 10918 E. 41st, 627-5000

The Westin Hotel Williams Center, 10 E.
2nd, 582-9000

SELECTED RESTAURANTS:

Charlie Mitchells, 81st St. & Lewis,
481-3250
Fountains Restaurant, 6540 S. Lewis
Ave., 749-9915
Interurban Restaurant, 717 S. Houston,
585-3134
Italian Inn, 5800 S. Lewis, 742-8886
Montagues, in The Westin Hotel Williams
Center, 582-9000
The Olive Garden, 7019 S. Memorial Dr.,
254-0082
Rosie's Rib Joint, 8125 E. 49th St.,
663-2610

SELECTED ATTRACTIONS:

Allen Ranch, 196th & S. Memorial,
366-3010
Bell's Amusement Park, 21st & New
Haven, 744-1991
Creek Nation Bingo, 1616 E. 81st St.,
299-0100
Discoveryland, 2502 E. 71st St., 496-0190
Gilcrease Museum, 1400 Gilcrease Rd.,
582-3122
Oral Roberts University campus, 7777 S.
Lewis, 495-6807
Perryman Wrangler Ranch, 11524 S.
Elwood, 299-2997
Philbrook Museum of Art, 2727 S.
Rockford, 749-7941
Tulsa Rose Garden/Woodward Park, 23rd
& Peoria, 749-6401
Tulsa Zoo, 5701 E. 36th St. N., 596-2400

INFORMATION SOURCE:

Convention & Visitors Bureau,
Metropolitan Tulsa Chamber of
Commerce
616 S. Boston Ave.
Tulsa, Oklahoma 74119
(918) 585-1201

Vancouver, British Columbia, Canada

Population: (*1,380,729)
414,281 (1986C)
Altitude: Sea level to 40 ft.
Average Temp.: Jan., 36.5°F.; July, 72°F.
Telephone Area Code: 604
Time: (none) **Weather:** 664-9032
Time Zone: Pacific

AIRPORT TRANSPORTATION:

Eleven miles to downtown Vancouver.
Taxicab and bus service.

SELECTED HOTELS:

Coast Plaza Hotel, 1733 Comox St.,
688-7711
Hotel Georgia, 801 W. Georgia St.,
682-5566
Hotel Vancouver, 900 W. Georgia St.,
684-3131
Hyatt Regency, 655 Burrard St., 687-6543
Pan Pacific Hotel, 999 Canada Pl.,
662-8111
Waterfront Centre Hotel, 900 Canada
Place Way, 691-1991
Westin Bayshore, 1601 W. Georgia St.,
682-3377

SELECTED RESTAURANTS:

The Cannery, 2205 Commissioner St.,
254-9606
Captain's Palace, 309 Belleville St.,
388-9191
Cavalier Room, in the Hotel Georgia,
682-5566
Hy's Encore, 637 Hornby St., 683-7671
Monk McQueens Fresh Seafood & Oyster
Bar, 601 Stamps Landing, 877-1351
Mulvaney's Restaurant, 1535 Johnston
St., on Granville Island, 685-6571
1066 Restaurant, 1066 Hastings St.,
689-1066

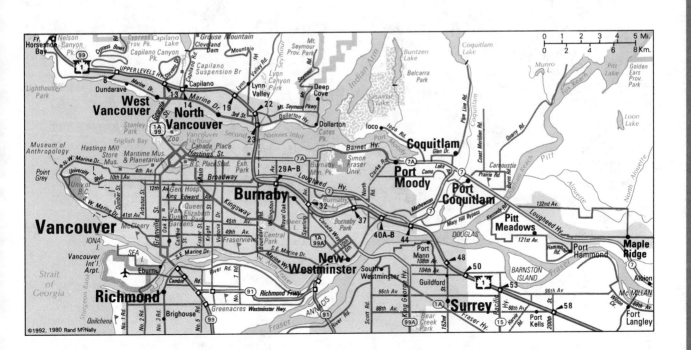

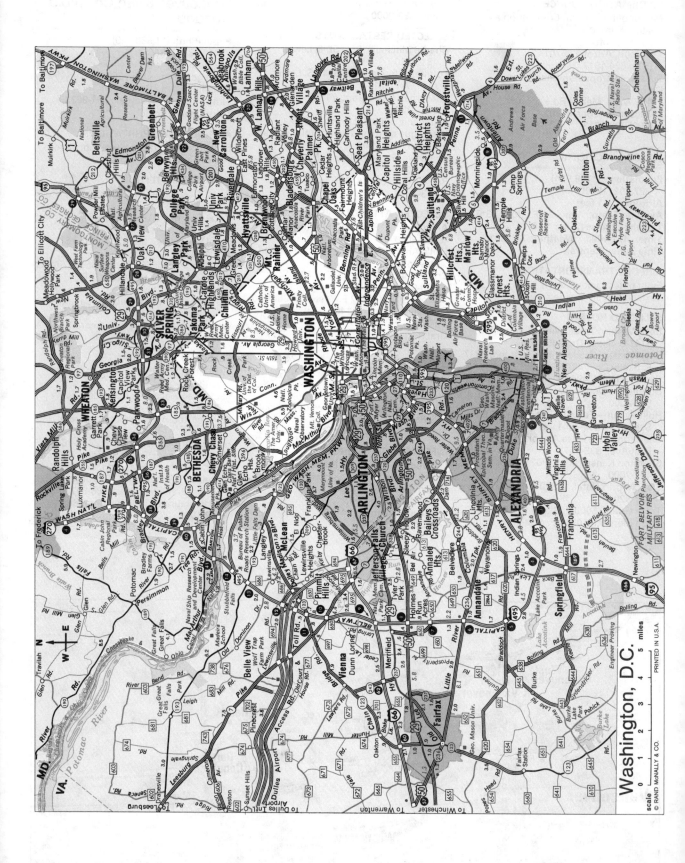

Washington, D.C.

scale
0   1   2   3   4   5   miles

© RAND McNALLY & CO.   PRINTED IN U.S.A.

The Three Greenhorns, in Denman Place
Mall, entrance on Comox, 688-8655
Umberto al Porto, 321 Water St.,
683-8376

SELECTED ATTRACTIONS:

Chinatown (includes the Dr. Sun Yat-Sen
Classical Chinese Garden), located
along Pender St. on both sides of Main
St.
Gastown (1880s district; includes the
Gastown Steam Clock), bounded by
Water, Carrall, Cordova, & Powell sts.
Granville Island, underneath the south
end of Granville St. bridge, 666-5784
Grouse Mountain (skiing, all-year
recreation), 6400 Nancy Greene Way,
North Vancouver, 984-0661
Museum of Anthropology, 6393 NW
Marine Dr., University of British
Columbia, 822-5087
Planetarium, 1100 Chestnut St., 736-3656
Robsonstrasse (European district),
Robson St. between Howe &
Broughton sts.
Science World, Quebec St. & Terminal
Ave., 687-7832
Vancouver Maritime Museum/St. Roch
National Historic Site, 1905 Ogden
Ave., 737-2211
Vancouver Public Aquarium, located in
Stanley Park (which includes numerous
other major attractions), 682-1118

INFORMATION SOURCES:

Greater Vancouver Convention & Visitors
Bureau
555 Burrard St., Suite 665
Vancouver, British Columbia V7X 1M8
(604) 682-2222
Vancouver Travel Info Centre
1055 Dunsmuir St.
P.O. Box 49296
Vancouver, British Columbia V7X 1L3
(604) 683-2000
Canadian Consulate General
1251 Avenue of the Americas
New York, New York 10020
(212) 768-2400

Washington, D.C.

Population: (*3,923,574)
606,900 (1990C)
Altitude: 1 to 410 feet
Average Temp.: Jan., 36°F.; July, 79°F.
Telephone Area Code: 202
Time: 844-2525   **Weather:** 936-1212
Time Zone: Eastern

AIRPORT TRANSPORTATION:

Three miles from Washington National
Airport to downtown Washington; 26
miles from Dulles International Airport
to downtown Washington; 37 miles
from Baltimore-Washington
International Airport to downtown
Washington.
Taxicab, bus, and rapid service between
National Airport and downtown
Washington; taxicab and bus service to
Dulles International Airport; taxicab,
bus, and train service to
Baltimore-Washington International
Airport.

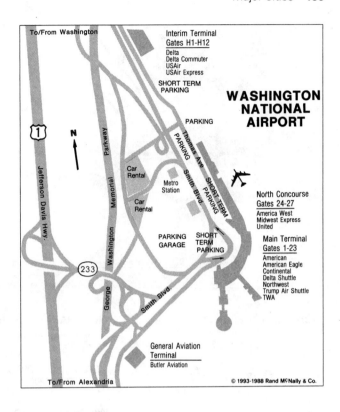

SELECTED HOTELS:

Capital Hilton, 16th & K sts. NW, 393-1000
Four Seasons Hotel, 2800 Pennsylvania
Ave. NW, 342-0444
Georgetown Inn, 1310 Wisconsin Ave.
NW, 333-8900
Grand Hyatt Washington, 1000 H St. NW,
582-1234
The Hay-Adams Hotel, 800 16th St. NW,
638-6600

Loew's L'Enfant Plaza, 480 L'Enfant Plaza
SW, 484-1000
Madison Hotel, 1177 15th St. NW,
862-1600
One Washington Circle Hotel, One
Washington Circle NW, 872-1680
The Ritz-Carlton, 2100 Massachusetts
Ave. NW, 293-2100
Sheraton-Carlton, 923 16th St. NW,
638-2626

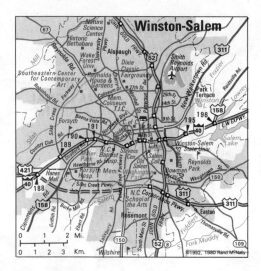

Washington Hilton and Towers, 1919
    Connecticut Ave. NW, 483-3000
The Watergate Hotel, 2650 Virginia Ave.
    NW, 965-2300

SELECTED RESTAURANTS:

Aux Beau Champs, in the Four Seasons
    Hotel, 342-0444
Cantina D'Italia Ristorante, 3251 Prospect
    St., 659-1830
Jockey Club Restaurant, in The
    Ritz-Carlton, 293-2100
Le Lion D'or, 1150 Connecticut Ave. NW,
    18th St. entrance, 296-7972
Maison Blanche Restaurant, 1725 F St.
    NW, 842-0070
Montpelier, in the Madison Hotel,
    862-1600
1789 Restaurant, 1226 36th St. NW,
    965-1789
Trader Vic's, in the Capital Hilton,
    347-7100

SELECTED ATTRACTIONS:

The Jefferson Memorial, Tidal Basin
    (South Bank), East Potomac Park,
    426-6822
The Lincoln Memorial, 23rd St. SW &
    Independence Ave., 426-6841
National Air & Space Museum, 6th St. &
    Indepedence Ave. SW, 357-2700
National Gallery of Art, 4th St. &
    Constitution Ave. NW, 737-4215
National Museum of American History,
    14th St. & Constitution Ave. NW,
    357-2700
National Museum of Natural History,
    10th St. & Constitution Ave. NW,
    357-2700
United States Capitol, National Mall (East
    End), 225-6827
Vietnam Veterans Memorial, Constitution
    Ave. & 22nd St. NW, 426-6841
The Washington Monument, National
    Mall at 15th St. NW & Constitution
    Ave., 426-6841
The White House, 1600 Pennsylvania
    Ave. NW, 456-7041

INFORMATION SOURCE:

Washington Convention and Visitors
Association
    1212 New York Ave. NW
    Washington, D.C. 20005
    (202) 789-7000

Winston-Salem, North Carolina

Population: (*942,091)
    (Greensboro-Winston Salem-High
    Point Metro Area)
    143,485 (1990C)
Altitude: 912 feet
Average Temp.: Jan., 41°F.; July, 78°F.
Telephone Area Code: 919
Time: 773-0000 **Weather:** 761-8411
Time Zone: Eastern

AIRPORT TRANSPORTATION:

Nineteen miles to downtown
    Winston-Salem.
Taxicab, bus and limousine service.

SELECTED HOTELS:

Marque Winston-Salem, 460 N. Cherry
    St., 725-1234
Ramada Inn North—Airport, 531 Akron
    Dr., 767-8240
Salem Inn, 127 S. Cherry St., 725-8561
The Stouffer Winston Plaza, 425 N.
    Cherry St., 725-3500
Tanglewood Park Lodge, U.S. 158, West
    Clemmons, 766-0591

SELECTED RESTAURANTS:

The Dirtwater Fox, in the Marque
    Winston-Salem, 725-1234
La Chaudière, 120 Reynolda Village,
    748-0269
Old Salem Tavern, 736 S. Main St.,
    748-8585
The Quill Restaurant, in the Stouffer
    Winston Plaza, 725-3500
Ryan's, 719 Coliseum Dr., 724-6132

SELECTED ATTRACTIONS:

Historic Bethabara Park, 2147 Bethabara
    Rd., 924-8191
Museum of Early Southern Decorative
    Arts, 924 S. Main St., 721-7360
Nature Science Center, I-52 at Hanes Mill
    Rd., 767-6730
Old Salem, S. Main St. & Academy,
    721-7300
Piedmont Craftsmen, 1204 Reynolda Rd.,
    725-1516
R.J. Reynolds Tobacco Company,
    Whitaker Park/Reynolds Blvd.,
    741-5718
Reynolda House Museum of American
    Art, 2250 Reynolda Rd., 725-5325
Southeastern Center for Contemporary
    Art, 750 Marguerite Dr., 725-1904
Stroh Brewery, 4791 Schlitz Ave.,
    788-6710
Tanglewood Park, US 158 West,
    Clemmons, 766-0591

INFORMATION SOURCES:

Winston-Salem Convention and Visitors
Bureau
    500 W. 5th St.
    P.O. Box 1408
    Winston-Salem, North Carolina 27102
    (919) 725-2361; (800) 331-7018
Winston-Salem Visitor Center
    601 N. Cherry St., Suite 100
    Winston-Salem, North Carolina 27101
    (919) 777-3796; (800) 331-7018
Greater Winston-Salem Chamber of
Commerce
    500 W. 5th St.
    P.O. Box 1408
    Winston-Salem, North Carolina 27102
    (919) 725-2361